Medically Speaking

A Dictionary of Quotations
on Dentistry, Medicine and Nursing

About the Compilers

Carl C Gaither was born on 3 June 1944 in San Antonio, Texas. He has conducted research work for the Texas Department of Corrections, the Louisiana Department of Corrections, and taught mathematics, probability, and statistics at McNeese State University and Troy State University at Dothan. Additionally he worked for ten years as an Operations Research Analyst. He received his undergraduate degree (Psychology) from the University of Hawaii and has graduate degrees from McNeese State University (Psychology), North East Louisiana University (Criminal Justice), and the University of Southwestern Louisiana (Mathematical Statistics).

Alma E Cavazos-Gaither was born on 6 January 1955 in San Juan, Texas. San Juan has the name of a big city but in Texas it's just a small border town. She has previously worked in quality control, material control, and as a bilingual data collector. In addition to compiling the quotations for this science quotation book series she is also an SK1 in the United States Navy Reserve. She received her associate degree (Telecommunications) from Central Texas College and presently is working toward a BA degree with a major in Spanish and a minor in Business Management.

Together they selected and arranged quotations for the books *Statistically Speaking: A Dictionary of Quotations* (Institute of Physics Publishing, 1996), *Physically Speaking: A Dictionary of Quotations on Physics and Astronomy* (Institute of Physics Publishing, 1997), *Mathematically Speaking: A Dictionary of Quotations* (Institute of Physics Publishing, 1998) and *Practically Speaking: A Dictionary of Quotations on Engineering, Technology and Architecture* (Institute of Physics Publishing, 1999).

About the Illustrator

Andrew Slocombe was born in Bristol in 1955. He spent four years of his life at Art College where he attained his Honours Degree (Graphic Design). Since then he has tried to see the funny side to everything and considers that seeing the funny side to medicine, nursing and dentistry has tested him to the full! He would like to thank Carl and Alma for the challenge!

Medically Speaking
A Dictionary of Quotations on Dentistry, Medicine and Nursing

Selected and Arranged by

Carl C Gaither
and
Alma E Cavazos-Gaither

Illustrated by Andrew Slocombe

Institute of Physics Publishing
Bristol and Philadelphia

IOP Publishing Ltd has attempted to trace the copyright holders of all the quotations reproduced in this publication and apologizes to copyright holders if permission to publish in this form has not been obtained.

British Library Cataloguing-in-Publication Data
A catalogue record for this book is available from the British Library.

ISBN 0 7503 0635 1

Library of Congress Cataloging-in-Publication Data are available

Production Editor: Al Troyano
Production Control: Sarah Plenty and Jenny Troyano
Commissioning Editor: Jim Revill
Editorial Assistant: Victoria Le Billon
Cover Design: Jeremy Stephens
Marketing Executive: Colin Fenton

Published by Institute of Physics Publishing, wholly owned by The Institute of Physics, London

Institute of Physics Publishing, Dirac House, Temple Back, Bristol BS1 6BE, UK
US Office: Institute of Physics Publishing, Suite 1035, The Public Ledger Building, 150 South Independence Mall West, Philadelphia, PA 19106, USA

Typeset in TeX using the IOP Bookmaker Macros
Printed in Great Britain by J W Arrowsmith Ltd, Bristol

This book is dedicated to my mother
Pearl Gaither, R.N.
and to my son
Russell J. Gaither, EMT

Carl C. Gaither

This book is dedicated to my sister
Rosie Cervantes, LPN (1952–1997)

Also I dedicate this book to my mother,
Magdelana Cavazos, who had to use every home remedy
known to mankind to keep all eleven of her
children in good health

I also dedicate this book to my grandson
Malcolm Xavier Childs

Alma E. Cavazos-Gaither

In memorium
In loving memory of Ethel Bernal
(9 November 1940–20 April 1999)
Wife, mother, sister and aunt

CONTENTS

PREFACE

Each year a large number of students enter the schools of dentistry, medicine, and nursing throughout the world. They enter what is to them a new world of thought and ideas. Even though advisors do suggest they read the classical works, the students usually don't have the time to delve deeply into these works. For the student *Medically Speaking* will provide a convenient way to quickly locate some of the great and not-so-great thoughts which have been written.

Rachel Carson wrote: *'I have, I confess, rather strong and definite prejudice against altering an author's words when excerpts from his writings are reprinted. A quotation, in my probably old-fashioned view, should be a quotation'* (from Paul Brooks *The House of Life: Rachel Carson at Work*, The Writer and His Subject, p. 3).

The aim of *Medically Speaking*, which contains over 1500 quotations, has been to provide all classes of medical people, as well as the non-practitioner who has an interest in medicine, with a volume of unaltered quotations. Another aim has been to provide a book that is attractive in appearance and of convenient size so that it may be kept on a desk for constant reference.

While there are other books of medical quotations, *Medically Speaking* has several important points of originality. Firstly, it has been freshly written 'from scratch' to give the widest possible range of quotations from the works of professionals (in and out of the field of medicine), poets, philosophers, writers, and anyone else we found who had said something worth repeating. As such, it is a work that has not appeared in print before. Secondly, it has illustrations. These illustrations have been included to bring a smile by showing to the eye a humorous visual interpretation of some of the quotations. Thirdly, it is worth pointing out that never before has so comprehensive a book of medical quotations been generally available to the public at so low a price as is *Medically Speaking*.

Quite a few of the quotations have been used frequently and will be recognized while others have probably not been used before. The authority for each quotation has been given with the fullest possible information that we could find so as to help you pinpoint the quotation in its appropriate context or discover more quotations in the original source. When the original source could not be located we indicated where we found the quote. Sometimes, however, we only had the quote and not the source. When this happened we listed the source as unknown and included the quotation anyway so that it would not become lost in time.

In summary, *Medically Speaking* is a book that has many uses. You can:

- Identify the author of a quotation.
- Identify the source of the quotation.
- Check the precise wording of a quotation.
- Discover what an individual has said on a subject.
- Find sayings by other individuals on the same subject.

How to Use This Book

1. A quotation for a given subject may be found by looking for that subject in the alphabetical arrangement of the book itself. To illustrate, if a quotation on "brain" is wanted, you will find seven quotations listed under the heading BRAIN. The arrangement of quotations in this book under each subject heading constitutes a collective composition that incorporates the sayings of a range of people.
2. To find all the quotations pertaining to a subject and the individuals quoted use the SUBJECT BY AUTHOR INDEX. This index will help guide you to the specific statement that is sought. A brief extract of each quotation is included in this index.
3. If you recall the name appearing in the attribution or if you wish to read all of an individual author's contributions that are included in this book then you will want to use the AUTHOR BY SUBJECT INDEX. Here the authors are listed alphabetically along with their quotations. The birth and death dates are provided for the authors whenever we could determine them.

Thanks

It is never superfluous to say thanks where thanks are due. Firstly, we want to thank Jim Revill and Al Troyano, of Institute of Physics Publishing, who have assisted us so very much with our books. Next, we thank the following libraries for allowing us to use their resources: The Jesse H. Jones Library and the Moody Memorial Library, Baylor University; the main library of the University of Mary-Hardin Baylor; the main library of the Central Texas College; the Undergraduate Library, the Engineering Library, the Law Library, the Physics-Math-Astronomy Library, and the Humanities Research Center, all of the University of Texas at Austin. Again, we wish to thank Joe Gonzalez, Chris Braun, Ken McFarland, Kathryn Kenefick, Gabriel Alvarado, Janice Duff, Rennison Lalgee, Deidra Allen, Brian Camp, Robert Clontz, Michelle Gonzales,

Katie MacInnis, Mike Harris, Brigid Spackman, Alex Marshall, Sammie Morris, and Ethan Perry of the Perry-Castañeda Library for putting up with us when we were checking out the hundreds of books. Finally, we wish to thank our children Maritza, Maurice, and Marilynn for their assistance in finding the books we needed when we were at the libraries.

A great amount of work goes into the preparation of any book. When the book is finished there is then time for the editors and authors to enjoy what they have written. It is hoped that this book will stimulate your imagination and interests in matters about dentistry, medicine and nursing. This objective has been expressed by Helen Hill (quoted in Llewellyn Nathaniel Edwards *A Record of History and Evolution of Early American Bridges*, p. xii):

> If what we have within our book
>   Can to the reader pleasure lend,
> We have accomplished what we wished,
>   Our means have gained our end.

Carl Gaither
Alma Cavazos-Gaither
August 1999

ABDOMEN

Unknown
The part of the body responsible for converting processed food into processed tissue.

In Richard Iannelli
The Devil's New Dictionary

ABORTION

Given, William P.
After it became legal, I tried performing them for a while. But when I'd get home I'd feel rotten. And yet I absolutely feel it's a woman's right. So now if a patient wants one I refer her to someone else, someone I know is skilled and reasonably priced. Does that make me a hypocrite?

In Louise Kapp Howe
Moments on Maple Street
Chapter Three (p. 21)

Hachamovitch, Moshe
By and large, legal or not, the procedure is still a pariah of our specialty.

In Louise Kapp Howe
Moments on Maple Street
Chapter Three (p. 21)

Kennedy, Flo
If men could get pregnant, abortion would be a sacrament.

In Roz Warren
Glibquips (p. 2)

Nolan, James Joseph
Physicians roasted on the spit;
Is learning the name of it?
For complications, spare no precaution;
To save a life think abortion.

The New England Journal of Medicine
On Renewed Maternal Mortality Reports (p. 952)
Volume 286, Number 17, April 27, 1972

Pope Pius XI
However we may pity the mother whose health and even life is imperiled by the performance of her natural duty, there yet remains no sufficient reason for condoning the direct murder of the innocent.

Casti Connubii
December 31, 1930

Tertullian

It's a committing murther before hand, to destroy that which is to be born . . .

Apologeticus
IX, 197

Unknown

Why not outlaw heterosexuality instead of abortion? Strike at the source!

Source unknown

ACUPUNCTURE

Unknown

A medical practice, employing needles, which offers relief from pain but no backing out. Once the patient agrees to the treatment, he's stuck with it.

In Richard Iannelli
The Devil's New Dictionary

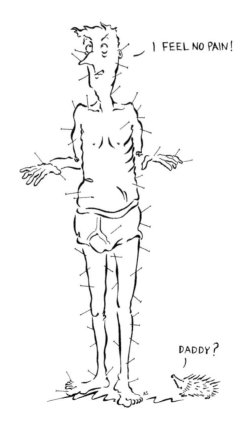

ADHESIVE

Armour, Richard
Removing adhesive is hazardous work:
Little by little? Or one sudden jerk?
Whichever it is, you may doubt you will win—
Removing adhesive, but leaving the skin.

The Medical Muse
Stuck with It

ALLERGY

Mather, Increase
Some men also have strange antipathies in their natures against that sort of food which others love and live upon. I have read of one that could not endure to eat either bread or flesh; of another that fell into a swooning fit at the smell of a rose . . .

Remarkable Providences
Chapter IV (p. 71)

Unknown
When a doctor doesn't know, he calls it a virus; when he does know but can't cure it, he calls it an allergy.

In Evan Esar
20,000 Quips and Quotes

Welsh, Joan I.
Medical science has gone far;
On that we'll all agree—
What used to be called an itch
Today's an allergy.

Quote, The Weekly Digest
July 21, 1968 (p. 56)

AMNESIA

Unknown
An affliction, usually caused by a deep shock, trauma or a bump on the head, which renders a person unable to remember who he is. Most people don't know who they are in the first place, and are therefore immune.

In Richard Iannelli
The Devil's New Dictionary

AMPUTATION

Middleton, Thomas
I'll imitate the pities of old Surgeons
To this lost limb, who, ere they show their art,
Cast one asleep, then cut the diseas'd part.

Women Beware Women
Act IV, Scene I (p. 91)

Webster, John
I had a limb corrupted to an ulcer,
But I have cut it off; and now I'll go
Weeping to heaven on crutches.

The White Devil
Act IV, Scene II, L. 117–119

ANATOMISTS

Richardson, Samuel
And I believe that anatomists allow *that women have more watery heads than men.*

The Works of Samuel Richardson
Volume VII
The History of Clarissa Harlowe
Volume IV
Letter XXVII (p. 130)

Twain, Mark
Surgeons and anatomists see no beautiful women in all their lives, but only a ghastly sack of bones with Latin names to them, and a network of nerves and muscles and tissues inflamed by disease.

Letter to the *Alta Californian*
San Francisco, May 28, 1867

ANATOMY

Bacon, Francis
In the inquiry which is made by anatomy, I find much deficience: for they inquire of the parts, and their substances, figures, and collocations; but they inquire not of the diversities of the parts, the secrecies of the passages, and the seats or nestling of the humours, nor much of footsteps and impressions of diseases.

Advancement of Learning
Second Book, X, 5

Burton, Robert
[Diseases] crucify the soul of man, attenuate our bodies, dry them, wither them, shrivel them up like old apples, make them as so many anatomies.

The Anatomy of Melancholy
Part I, Section 2, Memb. 3, Subsection 10

Dagi, Teodoro Forcht
Ask any doctor off the street
To speak of his most prizèd feat:
No doubt he'd answer honestly,
And say "to pass anatomy".

The New England Journal of Medicine
Anatomy of the Brain and Spinal Medulla: A Manual for Students (p. 1010)
Volume 286, Number 18, May 4, 1972

Dickinson, Emily
A science—so the Savants say,
"Comparative Anatomy"—
By which a single bone—
Is made a secret to unfold
Of some rare tenant of the mold,
Else perished in the stone—

The Complete Poems of Emily Dickinson
#3

Fernel, Jean
Anatomy is to physiology as geography is to history; it describes the theater of events.

De Naturali Parte Medicinae Libri Septem
Chapter I

Halle, John
But chieflye the anatomye
Ye oughte to understande:
If ye will cure well anye thinge,
That ye doe take in hande.

In Mary Lou McDonough
Poet Physician
Anatomy (p. 11)

Holmes, Oliver Wendell
What geology has done for our knowledge of the earth, has been done for our knowledge of the body by that method of study to which is given the name of *General Anatomy*.

Medical Essays
Border Lines in Medical Science (p. 222)

Muller, Herbert J.
To say . . . that a man is made up of certain chemical elements is a satisfactory description only for those who intend to use him as a fertilizer.

Science and Criticism
Chapter V (p. 107)

Nye, Bill
The word anatomy is derived from two Greek spatters and three polywogs, which, when translated, signify "up through" and "to cut", so that anatomy actually, when translated from the original wappy-jawed Greek, means to cut up through. That is no doubt the reason why the medical student proceeds to cut up through the entire course.

Remarks
Anatomy (p. 27)

Human anatomy is either general, specific, topographical or surgical. These terms do not imply the dissection and anatomy of generals, specialists, topographers and surgeons, as they might seem to imply, but really mean something else. I would explain here what they actually do mean if I had more room and knew enough to do it.

Remarks
Anatomy (p. 28)

Osler, Sir William

Anatomy may be likened to a harvest-field. First come the reapers, who, entering upon untrodden ground, cut down a great store of corn from all sides of them. These are the early anatomists of modern Europe, such as Vesalius, Fallopius, Malpighi, and Harvey. Then come the gleaners, who gather up ears enough from the bare ridges to make a few loaves of bread. Such were the anatomists of the last century—Valsalva, Cotunnius, Haller, Winslow, Vicq d'Azyr, Camper, Hunter, and the two Monroes. Last of all come the geese, who still contrive to pick up a few grains scattered here and there among the stubble, and waddle home in the evening, poor things, cackling with joy because of their success. Gentlemen, we are the geese.

Aequanimitas
The Leaven of Science
II (pp. 84–5)

Reid, Thomas

If a thousand of the greatest wits that ever the world produced were, without any previous knowledge in anatomy, to sit down and contrive how, and by what internal organs, the various functions of the human body are carried on, how the blood is made to circulate and the limbs to move, they would not, in a thousand years, hit upon anything like the truth.

The Works of Thomas Reid
Essays on the Intellectual Powers of Man
Essay I, Chapter III (p. 235)

Shapp, Paul

The human body comes in only two shapes and three colors. I don't expect there will be any changes, so what we learn about it now will serve us for a long time to come.

Time
The Fastest Man on Earth (p. 88)
Volume LXVI, Number 11, September 12, 1955

ANESTHESIA

Armour, Richard
Behold the patient uncomplaining,
Not asking whether losing, gaining,
Not offering unsought advice,
But really being very nice.

. . .

Behold the patient quite relaxed,
With nerves, this once, not overtaxed,
Serene, almost unrecognized,
Not fighting back—anesthetized.

The Medical Muse
Behold the Patient

Du Bartas, Guillaume de Saluste
Even as a Surgeon, minding off-to-cut
Some cure-less Limb; before in ure he put
His violent Engines on the vicious member,
Bringeth his Patient in a sense-less slumber,
And grief-less then (guided by Life and Art),
To save the whole; saws off th' infested part; . . .

Du Bartas His Divine Weekes and Works
First Week, Sixth Day (p. 57)

Genesis 2:21
And the Lord God caused a deep sleep to fall upon Adam, and he slept:
and he took one of his ribs, and closed up the flesh instead thereof.

The Bible

Helmuth, William Tod
For thus we read (although the analgesia
Of Richardson was then entirely unknown)

13

Adam profoundly slept with anaesthesia,
And from *his thorax was removed a bone*.
This was the first recorded operation,
(No doctor here dare tell me that I fib!)
And surgery, thus early in creation,
Can claim complete excision of a rib!

> *Scratches of a Surgeon*
> Surgery vs. Medicine (p. 66)

Holmes, Oliver Wendell

. . . three natural anaesthetics—sleep, fainting, death . . .

> *Medical Essays*
> The Medical Profession in Massachusetts (p. 365)

Kraus, Karl

Anesthesia: wounds without pain.

> *Half-Truths & One-and-a-Half Truths* (p. 112)

Massinger, Philip

1 Doct. We have given her, sir,
A sleepy potion, that will hold her long,
That she may be less sensible of the torment
The searching of the wound will put her to.

> *The Plays of Philip Massinger*
> Volume I
> The Duke of Milan
> Act V, Scene II (p. 337)

Unknown

NOVOCAIN: An anesthesia that helps to deaden the pain of the music in a dentist's office.

> In Richard Iannelli
> *The Devil's New Dictionary*

ANESTHETIST

Cvikota, Raymond J.
Anesthetist's cone: Ether bonnet.

Quote, The Weekly Digest
October 27, 1968 (p. 337)

Trotter, Wilfred
Mr. Anaesthetist, if the patient can keep awake, surely you can.

Attributed
The Lancet
Very Special Article (p. 1340)
Volume 2, 1965

Anesthesia: wounds without pain.
Karl Kraus – (See opposite)

ANTIBIOTIC

Unknown

[Antibiotic] What to give the man who has everything.

Esar's Comic Dictionary

APOTHECARY

Bierce, Ambrose
APOTHECARY, *n*. The physician's accomplice, undertaker's benefactor
and grave-worm's provider.

When Jove sent blessings to all men that are,
And Mercury conveyed them in a jar,
That friend of tricksters introduced by stealth
Disease for the apothecary's health,
Whose gratitude impelled him to proclaim:
"My deadliest drug shall bear my patron's name!"

The Enlarged Devil's Dictionary

Colman, George (the Younger)
A man, in a country town, we know,
    Professes openly with death to wrestle;
Ent'ring the field against the grimly foe,
    Armed with a mortar and a pestle.
Yet, some affirm, no enemies they are;
But meet just like prize-fighters, in a fair,
Who first shake hands before they box,
Then give each other plaguy knocks,
With all the love and kindness of a brother:
    So, many a suff'ring patient saith,
    Though the Apothecary fights with Death,
Still they're sworn friends to one another.

An Anthology of Humorous Verse
Selected by Helen & Lewis Melville
The Newcastle Apothecary

Hazlitt, William Carew
One said an Apothecaryes house must needs be healthful, because the windows, benches, boxes, and almost all the things in the house, tooke physick.

Shakespeare Jest Books
Volume III
Conceit, Clichés, Flashes and Whimzies
Number 41

Pope, Alexander
So modern *Pothecaries* taught the Art
By *Doctor's Bills* to play the *Doctor's Part*,
Bold in the Practice of *mistaken Rules*,
Prescribe, apply, and call their *Masters Fools*.

An Essay on Criticism
Part I, L. 108–111

Shakespeare, William
I do remember an apothecary—
And hereabouts he dwells,—which late I noted
In tatter'd weeds, with overwhelming brows,
Culling of simples; meager were his looks,
Sharp misery had worn him to the bones:
And in his needy shop a tortoise hung,
An alligator stuff'd, and other skins
Of ill-shaped fishes; and about his shelves
A beggarly account of empty boxes,
Green earthen pots, bladders and musty seeds,
Remnants of packthread and old cakes of roses,
Were thinly scatter'd to make up a show.

Romeo and Juliet
Act V, Scene I, L. 37–48

APPENDIX

Unknown

[Appendix] An internal organ of no value to anyone except a surgeon.

Esar's Comic Dictionary

[Appendix] Something that gives you information of inflammation.

Esar's Comic Dictionary

Appendicitis is caused by information in the appendix.

In Alexander Abingdon
Bigger & Better Boners (p. 72)

ARTERY

Barnes, Djuna
But the great doctor, he's a divine idiot and a wise man. He closes one eye, the eye that he studied with, and putting his finger on the arteries of the body says: 'God whose roadway this is, has given me permission to travel on it also,' which, Heaven help the patient, is true . . .

Nightwood
La Somnambule (p. 40)

BACK

Hubbard, Elbert
BACK: 2. A smooth surface composed of skin and bones which stretches
between Land's End and John O'Groat's.

The Roycroft Dictionary (p. 15)

BILL

da Costa, J. Chalmers
A fashionable surgeon, like a pelican, can be recognized by the size of his bill.

The Trials and Triumphs of the Surgeon
Chapter 1 (p. 38)

Hazlitt, William Carew
One asked a man whether he had swallowed a Doctor of Phisickes bill, because hee spoke such hard words.

Shakespeare Jest Books
Volume III
Conceit, Clichés, Flashes and Whimzies
Number 9

Morris, Robert Tuttle
One must not count upon all of his patients being willing to steal in order to pay doctor's bills.

Doctors versus Folks
Chapter 3

Unknown
The doctor cures all kinds of ills,
Except the shock of doctor's bills.

Source unknown

BIRTH CONTROL

Dickens, Charles
Accidents will occur in the best-regulated families.

The Works of Charles Dickens
David Copperfield
Volume II
Chapter 28 (p. 412)

Farris, Jean
Birth control: Banned parenthood.

Quote, The Weekly Digest
February 18, 1968 (p. 137)

Gabor, Dennis
The technique of birth control can be suppressed only if one abolishes also the technique of death control: medicine and hygiene.

Inventing the Future
Overpopulation (p. 82)

Sanger, Margaret
"Yes, yes—I know, Doctor", said the patient with trembling voice, "but," and she hesitated as if it took all of her courage to say it, "*what* can I do to prevent getting that way again?"

"Oh, ho!" laughed the doctor good naturedly. "You want your cake while you eat it too, do you? Well, it can't be done . . . I'll tell you the only sure thing to do. Tell Jake to sleep on the roof!"

My Fight for Birth Control
Awaking and Revolt (pp. 52–3)

The menace of another pregnancy hung like a sword over the head of every poor woman . . .

My Fight for Birth Control
Awaking and Revolt (p. 49)

No woman can call herself free who does not own and control her body.
No woman can call herself free until she can choose consciously whether
she will or will not be a mother.

Parade
December 1, 1963

Waugh, Evelyn
Impotence and sodomy are socially O.K. but birth control is flagrantly
middle-class.

In Nancy Mitford
Noblesse Oblige
'An Open Letter' Part III (p. 71)

Removing adhesive is hazardous work:
Little by little? Or one sudden jerk?...
Richard Armour – (See p. 5)

BLINDNESS

Keats, John
There is a budding morrow in midnight;
There is a triple sight in blindness keen; . . .

The Poems of John Keats
To Homer

Sophocles
Oedipus: As they say of the blind,
Sounds are the things I see

Oedipus at Colonus
Choral Dialogue (p. 13)

Thurber, James
Last night I dreamed of a small consolation enjoyed only by the blind:
Nobody knows the trouble I've *not* seen!

Newsweek
June 16, 1958

BLOOD

Armstrong, John
The blood, the fountain whence the spirits flow,
The generous stream that waters every part,
And motion, vigor, and warm life conveys
To every particle that moves or lives; . . .

Art of Preserving Health
Book II, L. 12–15 (p. 26)

Deuteronomy 12:23
The blood is the life.

The Bible

Du Bartas, Guillaume de Saluste
Even so the Blood (bred of good nourishment)
By divers Pipes to all the body sent,
Turns here to Bones there changes into Nerves;
Here is made Marrow, there for Muscles serves, . . .

Du Bartas His Divine Weekes and Works
First Week, Sixth Day (p. 55)

Editor of the Louisville Journal
"Doctor, what do you think is the cause of this frequent rush of blood to my head?"

"Oh, it is nothing but an effort of nature. Nature, you know, *abhors a vacuum*."

In George Denison Prentice
Prenticeana (p. 22)

Goethe, Johann Wolfgang von
MEPHISTOPHELES: Blood is a most peculiar fluid, my friend!

Faust
The First Part of the Tragedy
The Study (2) (p. 82)

Proverb
Blood is thicker than water.

Source unknown

Shakespeare, William
With purple fountains issuing from your veins.

Romeo and Juliet
Act I, Scene I, L. 92

BLOOD PRESSURE

Osler, Sir William

A man's life may be said to be a gift of his blood pressure, just as Egypt is a gift of the Nile.

In Harvey Cushing
The Life of Sir William Osler
Volume II (p. 297)

BODY

Butler, Samuel
The body is but a pair of pincers set over a bellows and a stewpan and the whole fixed upon stilts.

Samuel Butler's Notebooks (p. 289)

Chesterton, Gilbert Keith
The trouble about always trying to preserve the health of the body is that it is so difficult to do without destroying the health of the mind.

Come to Think of It
On the Classics (p. 47)

Flaubert, Gustave
Body. If we knew how our body is made, we wouldn't dare move.

Dictionary of Accepted Ideas

Heaney, Robert P.
It's just like remodeling an office . . . The body tears out partitions, puts up dry walls and paints.

Newsweek
The Calcium Craze (p. 50)
January 27, 1986

Heschel, Abraham J.
The body is a sanctuary, the doctor is a priest.

The Insecurity of Freedom
The Patient as a Person (p. 33)

Novalis, Friedrich Leopold
We touch heaven when we lay our hand on a human body.

In Robert Coope
The Quiet Art (p. 117)

Plato

. . . we are imprisoned in the body, like an oyster to his shell.

Phaedrus
[250] (p. 126)

Psalms 139:14

I am fearfully and wonderfully made.

The Bible

Romanoff, Alexis Lawrence

To man the human body is most sacred.

Encyclopedia of Thoughts
Aphorisms 1538

Unknown

Our body is a wonderful thing because without our body we would not live.

In Alexander Abingdon
Bigger & Better Boners (p. 68)

BODY-SNATCHER

Bierce, Ambrose

[Body-snatcher] One who supplies the young physicians with that which the old physicians have supplied the undertaker.

The Devil's Dictionary

BRAIN

Armstrong, John
. . . The secret mazy channels of the brain.

Art of Preserving Health
Book I, L. 178 (p. 11)

Day, Clarence
When the brain fails to act with the body, or, worse, works against it, the body will sicken no matter what cures doctors try.

This Simian World
Chapter Ten (p. 59)

Diamond, Marian
The brain is a three-pound mass you can hold in your hand that can conceive of a universe a hundred-billion light years across.

In John D. Barrow
Impossibility
Chapter 4
Complexity Matching (p. 96)

Huxley, Julian
The brain alone is not responsible for mind, even though it is a necessary organ for its manifestation. Indeed an isolated brain is a piece of biological nonsense.

In Pierre Teilhard de Chardin
The Phenomenon of Man
Introduction (pp. 16–7)

Langer, Susan
The brain works as naturally as the kidneys and the blood vessels. It is not dormant just because there is no conscious purpose to be served at the moment . . . Instead of that, it goes right on manufacturing ideas—streams and deluges of ideas, that the sleeper is not using to *think* with about anything. But the brain is following its own law; it is actively

translating experience into symbols, in fulfillment of a basic need to do so. It carries on a constant process of ideation.

Philosophy in a New Key (p. 46)

Ramón y Cajal, Santiago
Like the entomologist in pursuit of brightly coloured butterflies, my attention hunted, in the flower garden of the gray matter, cells with delicate and elegant forms, the mysterious butterflies of the soul, the beating of whose wings may some day—who knows?—clarify the secret of mental life.

Recollections of My Life
Chapter VII (p. 363)

Vogt, K.
The brain . . . is simply an organ which excretes feeling as the kidneys excrete urine.

In Irving John Good (Editor)
The Scientist Speculates
Mind and Consciousness (p. 80)

CANCER

Haldane, J.B.S.
I wish I had the voice of Homer
To sing of rectal Carcinoma,
Which kills a lot more chaps, in fact,
Than were bumped off when Troy was sacked.
I noticed I was passing blood
(Only a few drops, not a flood).
So passing on my homeward way
From Tallahassee to Bombay
I asked a doctor, now my friend,
To peer into my hinder end,
To prove or to disprove the rumor
That I had a malignant tumor.

New Statesman
Cancer's a Funny Thing (p. 298)
February 21, 1964

Mayo, Charles H.
While there are several chronic diseases more destructive to life than cancer none is more feared.

Annals of Surgery
Carcinoma of the Right Segment of the Colon
Volume 83, March 1926

Smithers, Sir David
Cancer is no more a disease of cells than a traffic jam is a disease of cars. A lifetime of study of the internal combustion engine would not help anyone to understand our traffic problems. A traffic jam is due to a failure in normal relationships between driven cars and their environment.

In Gothard Booth
The Cancer Epidemic: Shadow of the Conquest of Nature (p. 20)

Wells, H.G.
". . . the motive that will conquer cancer will not be pity nor horror; it will be curiosity to know how and why".

"And the desire for service," said Lord Tamar.

"As the justification of that curiosity," said Mr. Sempack, "but not as a motive. Pity never made a good doctor, love never made a good poet. Desire for service never made a discovery."

Meanwhile
Chapter 5 (p. 44)

CARE

Montaigne, Michel de

. . . the general order of things that takes care of fleas and moles, also takes care of men, if they will have the same patience that fleas and moles have, to leave it to itself.

Essays
Book the Second
Chapter 37 (p. 370)

Peabody, Francis Weld

The treatment of a disease may be entirely impersonal; the care of a patient must be completely personal.

The Care of the Patient (p. 12)

The secret of the care of the patient is in caring for the patient.

The Care of the Patient (p. 48)

CARTILAGE

Unknown
The cartilage is a liquid in the joints to keep the noise down.

In Alexander Abingdon
Bigger & Better Boners (p. 68)

CASTOR OIL

Crookshank, F.G.

. . . so, a dose of castor oil acts with equal efficiency whether given to expel a demon, to calm the vital spirits, to assuage the Archaeus, to evacuate morbific humours, to eliminate toxins, to restore endocrine balance, or to reduce blood pressure.

In C.G. Cumston
An Introduction to the History of Medicine from the Time of
the Pharaohs to the End of the XVIIIth Century (p. xxix)

CAVITY

Crest® toothpaste advertisement
Look, Mom—no cavities.

Davies, Robertson
I went to my dentist today in a high state of apprehension. None of my teeth were hurting me, but I know from bitter experience that a tooth of mine can have a cavity as big as the Grotto at Lourdes before it informs me of the fact.

The Table Talk of Samuel Marchbanks (p. 202)

Unknown
[Cavity] A hollow place in a tooth ready to be stuffed with a dentist's bill.

Esar's Comic Dictionary

Walis, Claudia
Tooth decay was a perennial national problem that meant a mouthful of silver for patients, and for dentists a pocketful of gold.

Time
Today's Dentistry: A New Drill (p. 73)
September 9, 1985

CHEMISTRY

Hoffmann, Roald
Carried by blood, carrying
electrons, life-empowering
oxygen. Elsewhere, in engines
it's sucked into carburetor

trains, there to mix with branched
heptanes, octanes, another kind
of feedstock. Sparked, it burns
things in controlled explosions,

a human specialty. And what
thermochemistry says should end
in greening CO_2 and steam, in
incomplete combustion partly

goes to CO, carbon monoxide.
This odorless diatomic tres-
passer sweeps into bronchia, brashly
binding 200 times better

than O_2. A free ride on deoxyhemo-
globin down arteries, right past cells that long for the other, can't
wait too long before shutdown.

Gaps and Verges
Jerry-Built Forever (p. 29)

Janssen, Johannes
. . . the doctor must be a chemist also, and medicine and chemistry cannot
be separated from each other.

History of the German People at the Close of the Middle Ages
Volume XIV
Chapter VI (p. 7)

Latham, Peter Mere
Sagacious observers and experimenters have, in these later days, gone nigh to show that there is a chemistry within us which is cooperative with life; that making good its work, it gives to our bodies the materials of their health; and that doing its work faultily, it suffers noxious things to form, which become the elements of their diseases.

In William B. Bean
Aphorisms from Latham (p. 89)

Let no man who is making his entrance into the medical profession henceforth ever neglect chemistry. Chemistry was once thought to be conversant only with the physiology of external nature; but every day is bringing us to look more and more to chemistry to explain the physiology of our own bodies.

In William B. Bean
Aphorisms from Latham (p. 89)

Osler, Sir William
. . . the physician without physiology and chemistry flounders along in an aimless fashion, never able to gain any accurate conception of disease, practising a sort of popgun pharmacy, hitting now the malady and again the patient, he himself not knowing which.

Aequanimitas
Teaching and Thinking
II (p. 121)

Shaw, George Bernard
In the arts of life man invents nothing; but in the arts of death he outdoes Nature herself, and produces by chemistry and machinery all the slaughter of plague, pestilence and famine.

Man and Superman
Act 3 (pp. 83–4)

CHILDBIRTH

Mitchell, Margaret

Death and taxes and childbirth! There's never any convenient time for any of them!

Gone with the Wind
Chapter xxxviii (p. 668)

Unknown

There was a young lady of York
Who was shortly expecting the stork,
  When the doctor walked in
  With a businesslike grin,
A pickax, a spade, and a fork.

In Louis Untermeyer
Lots of Limericks (p. 144)

[Childbirth] The outcome of an entrance.

Esar's Comic Dictionary

CHOLERA

Chekhov, Anton
He died from fear of cholera.

Note-Book of Anton Chekhov (p. 68)

Flaubert, Gustave
Cholera: You catch it from eating melons. The cure is lots of tea with rum in it.

Dictionary of Accepted Ideas

Inge, W.R.
If . . . an outbreak of cholera might be caused either by an infected water supply or by the blasphemies of an infidel mayor, medical research would be in confusion.

Outspoken Essays
Second Series
Confessio Fidei (p. 3)

Kipling, Rudyard
When the cholera comes—as it will past a doubt—
Keep out of the wet and don't go on the shout,
For the sickness gets in as the liquor dies out,
An' it crumples the young British soldier.
Crum-, crum-, crumples the soldier . . .

Collected Verse of Rudyard Kipling
The Young British Soldier

CIRCUMCISION

Freeland, E. Harding
It has been urged as an argument against the universal adoption of circumcision that the removal of the protective covering of the glans tends to dull the sensitivity of that exquisitely sensitive structure and thereby diminishes sexual appetite and the pleasurable effects of coitus. Granted that this be true, my answer is that, whatever may have been the case in days gone by, sensuality in our time needs neither whip nor spur, but would be all the better for a little more judicious use of curb and bearing-rein.

The Lancet
Circumcision as a Preventative of Syphilis and Other Disorders
Volume 2, December 29, 1900 (pp. 1869–71)

Johnson, Athol A.W.
In cases of masturbation we must, I believe, break the habit by inducing such a condition of the parts as will cause too much local suffering to allow of the practice to be continued. For this purpose, if the prepuce is long, we may circumcise the male patient with present and probably with future advantages; the operation, too, should not be performed under chloroform, so that the pain experienced may be associated with the habit we wish to eradicate.

The Lancet
On An Injurious Habit Occasionally Met with in Infancy and Early Childhood
Volume 1, April 7, 1860 (pp. 344–5)

Sayer, Lewis L.
Hip trouble is from falling down, an accident that children with tight foreskins are specially liable to, owing to the weakening of the muscles produced by the condition of the genitals.

Journal of the American Medical Association
Circumcision for the Cure of Enuresis
Volume 7, 1887 (pp. 631–3)

Taylor, A.W.
Not infrequently marital unhappiness would be better relieved by
circumcising the husband than by suing for divorce.

Medical Record
Circumcision—Its Moral and Physical Necessities and Advantages
Volume 56, 1899 (p. 174)

I think scientists have one thing in common with children...
Otto Frisch – (See p. 62)

COD LIVER OIL

Crichton-Browne, Sir James

Oleum Jecoris Aselli, or cod-liver oil, was introduced into medical practice in this country, by Dr., afterwards Professor John Hughes Bennett, of Edinburgh in 1841. It had long been used by the Laplanders as a delicacy, in the Shetland Islands in place of butter, and in Holland it had been recommended as a cure for gout and rheumatism, but its first employment as a medical agent amongst us followed on the publication of Dr. Bennett's treatise on the subject.

From the Doctor's Notebook
Oleum Jecoris Aselli (p. 39)

COLDS

Benchley, Robert
If you think you have caught a cold, call a good doctor. Call in three doctors and play bridge.

Benchley or Else
How to Avoid Colds (p. 166)

Chamfort, Nicolas
The threat of a neglected cold is for doctors what the threat of purgatory is for priests—a gold mine.

In Herbert V. Prochnow and Herbert V. Prochnow, Jr
A Treasury of Humorous Quotations
#1092

Crichton-Browne, Sir James
At a meeting of a medical society, it was said of the common cold that it was three days coming, three days staying, and three days going. A French physician, quoted by Sir St. Clair Thompson, said that the common cold, if left to itself, ran for a fortnight, but if medically treated, lasted only fourteen days.

The Doctor's After Thoughts (p. 235)

Simmons, Charles
The best way to cure a cold is, not to catch another.

Laconic Manual and Brief Remarker (p. 87)

Zealously nurse a cold with warm weather, and light and scanty food, till well cured, *or repentance will be upon you*.

Laconic Manual and Brief Remarker (p. 87)

Unknown
There was an old lady who said
When she found a thief under her bed,
   "Get up from the floor;
   You're too near the door,
And you may catch a cold in your head."

In Louis Untermeyer
Lots of Limericks (p. 72)

[Cold] A curious ailment that only people who are not doctors know how to cure.

Esar's Comic Dictionary

[Cold] An ailment for which there are many unsuccessful remedies, with whiskey being the most popular.

Esar's Comic Dictionary

[Cold] A respiratory ailment in which a virus attacks one's weakest point, which explains the prevalence of head colds.

Esar's Comic Dictionary

[Cold] A condition in the body marked by a runny nose, a sore throat, congested sinuses, and a resurgence of the childhood desire for a lot of mothering.

In Richard Iannelli
The Devil's New Dictionary

Young, Edward
I've known my lady (for she loves a tune)
For *fevers* take an opera in *June*
And perhaps you'll think the practice bold,
A midnight park is sovereign for a *cold*.

Love of Fame
Satire V, L. 185 (pp. 94, 95)

Wynne, Shirley W.
A person's age is not dependent upon the number of years that have passed over his head, but upon the number of colds that have passed through it.

Source unknown
Quoting Dr. Woods Hutchinson

COMMON SENSE

Arnauld, Antoine
Common sense is not really so common.

The Art of Thinking: Port-Royal Logic
First Discourse (p. 9)

Cross, Hardy
Common sense is only the application of theories which have grown and been formulated unconsciously as a result of experience.

Engineers and Ivory Towers
For Man's Use of God's Gifts (p. 107)

Einstein, Albert
. . . common sense is nothing more than a deposit of prejudices laid down in the mind before you reach eighteen.

In Eric T. Bell
Mathematics: Queen and Servant of Science (p. 42)

Goldenweiser, A.
. . . the physician would be even worse off than he is, if not for the occasional emergence of common sense which breaks through dogmas with intuitive freshness, or those flashes of insight for which talented diagnosticians are noted, or finally, an opportunity to make a biographical study of a patient, a luxury few physicians can enjoy and few patients can afford.

With the subject of the uniqueness of particulars, is ushered in intuitive mind as it functions in religion, art, and other forms of imaginative creativeness.

Robots or Gods (p. 62)

James, William
Common-sense contents itself with the unreconciled contradiction, laughs when it can, and weeps when it must, and makes, in short, a practical compromise, without trying a theoretical solution.

Collected Essays and Reviews
German Pessimism (p. 17)

Latham, Peter Mere
A small overweight of knowledge is often a sore impediment to the movement of common sense.

In William B. Bean
Aphorisms from Latham (p. 37)

Russell, Bertrand
Common sense, however it tries, cannot avoid being surprised from time to time. The aim of science is to save it from such surprises.

In Jean-Pierre Luminet
Black Holes (p. 182)

Whitehead, Alfred North
Now in creative thought common sense is a bad master. Its sole criterion for judgment is that the new ideas shall look like the old ones. In other words it can only act by suppressing originality.

An Introduction to Mathematics
Chapter 11 (p. 116)

CONSULTATION

Halle, John
When thou arte callde at anye time,
A patient to see;
And doste perceave the cure too grate,
And ponderous for thee:
See that thou laye disdeyne aside,
And pride of thyne owne skyll:
And thinke no shame counsell to take,
But rather wyth good wyll.
Gette one or two of experte men,
To helpe thee in that nede;
And make them partakers wyth thee,
In that worke to procede.

In Mary Lou McDonough
Poet Physician
Goodly Doctrine and Instruction (p. 11)

Holmes, Oliver Wendell
Now when a doctor's patients are perplexed,
A *Consultation* comes in order next—
You know what that is? In certain a certain place
Meet certain doctors to discuss a case
And other matters, such as weather, crops,
Potatoes, Pumpkins, lager-beer, and hops.
For what's the use!—there's little to be said,
Nine times in ten your man's as good as dead;
At best a talk (the secret to disclose)
Where three men guess and *sometimes* one man knows.

The Complete Poetical Works of Oliver Wendell Holmes
Rip Van Winkle, M.D.
Canto Second

CONTAGIOUS

Shakespeare, William
Will he steal out of his wholesome bed,
To dare the vile contagion of the night?

Julius Caesar
Act II, Scene I, L. 264–265

CONTRACEPTIVE

Allen, Woody

A fast word about oral contraception. I asked a girl to go to bed with me and she said 'no'.

Woody Allen Volume Two
Colpix CP. 488
Side 4, band 6

Chatton, Milton J.

Jack told Jill to take her pill
  With a glass of water.
Jill forgot, and Jack begot
  A bouncing baby daughter.

Quotable Quotes
March 13, 1966 (p. 16)

Glasser, Allen

When a patient asked which sulfa compounds make the safest contraceptives, his doctor replied:

"Sulfa-denial and sulfa-control!"

Quote, The Weekly Digest
May 7, 1967 (p. 377)

Sharpe, Tom

Skullion had little use for contraceptives at the best of times. Unnatural, he called them, and placed them in the lower social category of things along with elastic-sided boots and made-up bow-ties. Not the sort of attire for a gentleman.

Porterhouse Blue
Chapter 9 (p. 96)

53

Unknown

The best contraceptive is a glass of cold water: not before or after, but instead.

Source unknown

[Contraceptive] An anti-inflation measure.

Esar's Comic Dictionary

[Contraceptive] The choice offered a woman between perpetual virginity and perpetual pregnancy.

Esar's Comic Dictionary

CONVALESCENCE

Crichton-Browne, Sir James
Thank Heaven! the crisis,
The trouble past,
And the lingering illness
Is over at last.
And the fever called 'Living'
Is conquered at last.

And I rest so composedly
Now, in my bed,
That any beholder
Might fancy me dead.
Might start on beholding me,
Thinking me dead.

The Doctor Remembers
Convalescence (p. 222)

Lamb, Charles
How convalescence shrinks a man back to his pristine stature! where is now the space, which he occupied so lately, in his own, in the family's eye?

Essays of Elia
The Last Essays of Elia
The Convalescent (p. 333)

Shaw, George Bernard
LUBIN: I enjoy convalescence. It is the part that makes the illness worth while.

Back to Methuselah
Part II (pp. 66–7)

Unknown
Convalescence is the period when you are still sick after you get well.

In Evan Esar
20,000 Quips and Quotes

COUGH

Griffiths, Trevor
McBRAIN: Cough and the world coughs with you. Fart and you stand alone.

The Comedians
Act I (p. 17)

Ray, John
A dry cough is the trumpeter of death.

A Complete Collection of English Proverbs (p. 5)

Second World War Health Slogan
Coughs and sneezes spread diseases. Trap the germs in your handkerchief.

Source unknown

Wodehouse, P.G.
Jeeves coughed one soft, low, gentle cough like a sheep with a blade of grass stuck in its throat.

The Inimitable Jeeves
Chapter 13 (p. 139)

Wolcot, John
And, doctor, do you really think
That ass's milk I ought to drink?
'T would quite remove my cough, you say,
And drive my old complaints away.
It cured yourself—I grant it true;
But then—'t was mother's milk to you!

In William Davenport Adams
English Epigrams
To a Friend who Recommended Ass's Milk
cclxxxvi

CURE

Advertisement

I have discovered the natural system of cure for all diseases, habits, defects, failings, etc., without the use of deleterious and pernicious drugs or medicines. Being Scientific, it is absolutely safe, simple, painless, pleasant, rapid, and infallible. Diseases like hysteria, epilepsy, rheumatism, loss of memory, paralysis, insanity and mania; addiction to smoking, opium, drink, etc.; impotence, sterility, adultery, and the like can be radically cured duly by My System.

In Aldous Huxley
Jesting Pilate
India & Burma (p. 119)

Alexander, Franz

We now feel we can cure the patient without his fully understanding what made him sick. We are no longer so interested in peeling the onion as in changing it.

Quoted in
Time
Psychoanalysis Then and Now (p. 68)
May 19, 1961

Amiel, Henri-Frédéric

There is no curing a sick man who believes himself in health.

Amiel's Journal
February 6, 1877 (p. 305)

Baruch, Bernard

There are no such things as incurables, there are only things for which man has not found a cure.

Address to the President's Committee on Employment of
the Physically Handicapped
News report of May 1, 1954

Beaumont, Francis
Fletcher, John
THIERRY: We study satisfaction; must the cure
          Be worse than the disease?

Beaumont and Fletcher
Thierry and Theodoret
Act IV, Scene ii

Browne, Sir Thomas
. . . we all labour against our owne cure, for death is the cure of all
diseases.

Religio Medici
Part II, Section 9 (p. 93)

Burton, Robert
It is in vaine to speake of Cures, or thinke of remedies, until such time
as we have considered of the Causes . . .

The Anatomy of Melancholy
Volume I
Part I, Section 2 (p. 171)

Butler, Samuel
'Tis not amiss, ere ye're giv'n o'er,
To try one desp'rate med'cine more;
For where your case can be no worse,
The desp'rat is the wisest course.

The Poetical Works of Samuel Butler
Volume I
Epistle of Hudibras to Sidrophel, L. 5–8

Diseases of their own accord,
But cures come difficult and hard.

The Poetical Works of Samuel Butler
Volume II
The Weakness and Misery of Man, L. 82–83

Crabbe, George
Man yields to custom, as he bows to fate,
In all things rul'd—mind, body, and estate:
In pain, in sickness, we for cure apply
To them we know not, and we know not why: . . .

Tales in Verse
Tale III (p. 46)

Descartes, René
In truth, the most important thing for curing illnesses and maintaining health is good humor and joy.

In René Dubos & Jean-Paul Escande
Quest: Reflections on Medicine, Science, and Humanity
Chapter III (p. 59)

Esar, Evan
The man who doctors himself with the aid of medical books, runs the risk of dying of a typographical error.

20,000 Quips and Quotes

Time heals everything, but don't try sitting it out in a doctor's reception room.

20,000 Quips and Quotes

Hardy, Thomas
And ill it therefore suits
The mood of one of my high temperature
To pause inactive while await me means
Of desperate cure for these so desperate ills!

The Dynasts
Part First
Act IV, Scene III

Herrick, Robert
To an old soare a long cure must goe on; . . .

The Complete Poems of Robert Herrick
Volume III
Great Maladies, Long Medicine (p. 50)

Hitchcock, Sir Alfred
I have a perfect cure for a sore throat—cut it.

In Evan Esar
20,000 Quips and Quotes

Johnson, Samuel
When desp'rate Ills demand a Speedy Cure,
Distrust is Cowardice, and Prudence Folly.

Irene
Act IV, Scene I (p. 52)

Kipling, Rudyard
The cure for this ill is not to sit still,
    Or frowst with a book by the fire;
But to take a large hoe and a shovel also,
    And dig till you gently perspire.

Just So Stories
How the Camel got his Hump (p. 15)

Latham, Peter Mere
Let cure be looked upon as concerned with the disease as such, and
having little or no regard to the individual patient whom it befalls.

> In William B. Bean
> *Aphorisms from Latham* (p. 60)

Longfellow, Henry Wadsworth
Forth then issued Hiawatha,
Wandered eastward, wandered westward,
Teaching men the use of simple
And the antidotes for poisons,
And the cure of all diseases.
Thus was first made known to mortals
All the mystery of Medamin,
All the sacred art of healing.

> *The Song of Hiawatha*
> Hiawatha's Lamentation (pp. 208–9)

Ray, John
What cannot be *cured* must be endured.

> *A Complete Collection of English Proverbs* (p. 97)

A *disease* known, is half cured.

> *A Complete Collection of English Proverbs* (p. 100)

Romanoff, Alexis Lawrence
It is human nature to think that if a small dose helps, a double dose will
cure.

> *Encyclopedia of Thoughts*
> Aphorisms 329

One's desire to live is the best cure for many illnesses.

> *Encyclopedia of Thoughts*
> Aphorisms 370

Activity is the best cure for many ills of body and mind.

> *Encyclopedia of Thoughts*
> Aphorisms 2252

Shadwell, Thomas
RAYMUND: Well a desperate disease must have a desperate Cure . . .

> *The Complete Works of Thomas Shadwell*
> Volume I
> The Humorists
> Act IV (p. 237)

Shakespeare, William
Care is no cure, but rather corrosive
For things that are not to be remedied.

The First Part of King Henry VI
Act III, Scene III, L. 3–4

Szasz, Thomas
Masturbation: the primary sexual activity of mankind. In the nineteenth century it was a disease; in the twentieth, it's a cure.

The Second Sin
Sex (p. 10)

Tolstoy, Leo
What can doctors cure?

War and Peace
Book X, Chapter XXIX (p. 449)

Trudeau, Edward
Guérir quelquefois, soulager souvent, consoler toujours.
[To sometimes cure, often help, always console.]

In René Dubos & Jean-Paul Escande
Quest: Reflections on Medicine, Science, and Humanity
Chapter III (p. 56)

Unknown
How D.D. swaggers—M.D. rolls!
    I dub them both a brace of noddies:
Old D.D. takes the cure of souls,
    And M.D. takes the cure of bodies.
Between them both what treatment rare
    Our souls and bodies must endure!
One takes the cure without the care,
    T' other the care without the cure.

In William Davenport Adams
English Epigrams
On Parsons versus Doctors, ccxxxviii

One physician cures you of the colic; two physicians cure you of the medicine.

Journal of the American Medical Association
The Fixed Eruption (p. 765)
Volume 190, 1964

CURIOSITY

Day, Clarence
Creatures whose mainspring is curiosity will enjoy the accumulating of facts, far more than pausing at times to reflect on those facts.

This Simian World
Chapter Nine (p. 51)

Frisch, Otto
. . . I think scientists have one thing in common with children: curiosity. To be a good scientist you must have kept this trait of childhood, and perhaps it is not easy to keep just this one trait. A scientist *has* to be curious like a child; perhaps one can understand that there are other childish features he hasn't grown out of.

What Little I Remember (p. 86)

Selye, Hans
The true scientist thrives on curiosity . . .

From Dream to Discovery
Chapter 1 (p. 10)

Scientific curiosity can be satisfied much more easily by reading the publications of others than by working in the lab. It may take years to prove by experimentation what we can learn in the few minutes needed to read the published end result. So let us not fool ourselves; the driving force is hardly sheer curiosity.

From Dream to Discovery
Chapter 1 (p. 15)

DEATH

Allman, David
Life is precious to the old person. He is not interested merely in thoughts of yesterday's good life and tomorrow's path to the grave. He does not want his later years to be a sentence of solitary confinement in society. Nor does he want them to be a death watch.

Address to National Conference of Christian and Jews
The Brotherhood of Healing
February 12, 1958

Bassler, Thomas J.
Two out of every three deaths are premature; they are related to loafer's heart, smoker's lungs and drinker's liver.

In James Fixx
The Complete Book of Running
Chapter 1 (p. 4)

Brackenridge, Hugh Henry
When the patient is dead, it was the disease killed him, not the Doctor. Dead men tell no tales.

Modern Chivalry
Part II, Volume I, Chapter X (p. 378)

Browne, Sir Thomas
Men that looke no further than their outsides, thinke health an appertinance unto life, and quarrell with their constitutions for being sick; but I that have examined the parts of man, and know upon what tender filaments that Fabrik hangs, doe wonder that we are not always so; and considering the thousand dores that lead to death, doe thank my God that we can die but once.

Religio Medici
Part I, Section 44 (p. 57)

Chamfort, Nicolas

Living is an illness to which sleep provides relief every sixteen hours. It's a palliative. The remedy is death.

Maximes et Pensées
Chapter 2

Chekhov, Anton

When there is someone in a family who has long been ill, and hopelessly ill, there come terrible moments when all those close to him timidly, secretly, at the bottom of their hearts wish for his death, . . .

The Portable Chekhov
Peasants (p. 238)

Fechner, Gustav

[At death] we step into a still more free, quite new domain, which is yet not detached from the other, but rather encloses it in a wider circle . . .

Life After Death
Continued Existence of Ideas (p. 124)

Fielding, Henry

There is nothing more unjust than the vulgar opinion, by which physicians are misrepresented as friends to death. On the contrary, if the number of those who recover by physic could be opposed to that of the martyrs to it, the former would rather exceed the latter. Nay, some are so cautious on this head, that, to avoid a possibility of killing the patient, they abstain from all method of curing, and prescribe nothing but what can neither do good nor harm. I have heard some of these, with great gravity, deliver it as a maxim that Nature should be left to do her own work, while the physician stands by as it were to clap her on the back and encourage her when she doth well.

Tom Jones
Book II, Chapter 9 (p. 34)

Flatman, Thomas

When on my sick-bed I languish,
Full of sorrow, full of anguish;
Fainting, gasping, trembling, crying,
Panting, groaning, speechless, dying,
Methinks I hear some gentle spirit say,
Be not fearful, come away.

Source unknown

Fuller, Thomas

[The physician] . . . when he can keep life no longer in, he makes a fair and easy passage for it to go out.

In Harvey Cushing
The Life of Sir William Osler
Volume II (p. 299)

Gibran, Kahlil
For life and death are one, even as the river and the sea are one.

The Prophet
Death (p. 87)

Lewis, C.S.
How much better for us if *all* humans died in costly nursing homes amid doctors who lie, nurses who lie, friends who lie . . .

The Screwtape Letters
V (p. 32)

McCullers, Carson
Death is always the same, but each man dies in his own way.

Clock Without Hands
Chapter 1 (p. 1)

Nashe, Thomas
Adieu! farewell earth's bliss!
This world uncertain is:
Fond are life's lustful joys,
Death proves them all but toys.
None from his darts can fly:
I am sick, I must die—
   Lord, have mercy on us!

In Robert Coope
The Quiet Art (p. 155)

Osler, Sir William
We speak of death as the King of Terrors, yet how rarely does the act of dying appear to be painful, how rarely do we witness AGONY in the last few hours. Strict, indeed, is the fell sergeant in his arrest, but few feel the iron grip; the hard process of nature's law is for the most of us mercifully effected, and death, like birth, is 'but a sleep and a forgetting.'

In Harvey Cushing
The Life of Sir William Osler
Volume I (p. 294)

Pope, Alexander
But just disease to luxury succeeds,
And ev'ry death its own avenger breeds.

The Poems of Alexander Pope
Volume III
Essay on Man
Epis. iii, L. 165–166

Proverb, Scottish
Death defies the doctor.

A Complete Collection of English Proverbs (p. 283)

Rous, Francis
Now Death his servant Sickness forth hath sent . . .

Thule
The Second Book
Canto 4 (p. 103)

Seneca
Anyone can stop a man's life, but no one his death; a thousand doors
open on to it.

Phoenissae

Shakespeare, William
It is silliness to live when to live is torment; and then
have we a prescription to die when death is our
physician.

Othello
Act I, Scene III, L. 307–309

The patient dies while the physician sleeps.

The Rape of Lucrece
L. 909

He had rather
Groan so in perpetuity, than be cured
By the sure physician, death.

Cymbeline
Act V, Scene IV, L. 4–6

By medicine life may be prolonged, yet death
Will seize the doctor too.

Cymbeline
Act V, Scene V, L. 29–30

No cataplasm so rare,
Collected from all simples that have virtue
Under the moon, can save the thing from death.

Hamlet
Act IV, Scene VII, L. 144–146

PRINCE HENRY: It is too late: the life of all his blood
Is touch'd corruptibly, and his pure brain,
Which some suppose the soul's frail dwelling-house,
Doth by the idle comments that it makes
Foretell the ending of mortality.

. . .

O vanity of sickness! Fierce extremes
In their continuance will not feel themselves.
Death, having prey'd upon the outward parts,
Leaves them invisible; and his siege is now
Against the mind, the which he pricks and wounds
With many legions of strange fantasies,
Which, in their throng and press to that last hold,
Confound themselves. 'Tis strange that death should sing.
I am the cygnet to this pale faint swan
Who chants a doleful hymn to his own death,
And from the organ-pipe of frailty sings
His soul and body to their lasting rest.

. . .

KING JOHN: Ay, marry, now my soul hath elbow-room;
It would not out at windows or at doors.
There is so hot a summer in my bosom
That all my bowels crumble up to dust.
I am a scribbled form drawn with a pen
Upon a parchment, and against this fire
Do I shrink up.

The Life and Death of King John
Act V, Scene VII, L. 1–5, 14–25, 28–34

Shaw, George Bernard
Do away with death and you do away with the need for birth.

Misalliance
Parents and Children (ix)

Twain, Mark
Whoever has lived long enough to find out what life is, knows how deep a debt of gratitude we owe to Adam, the first great benefactor of our race. He brought death into the world.

The Tragedy of Pudd'nhead Wilson
Chapter III

Unknown
[Death] Nature's withdrawal from its battle with your doctor's treatment of your illness.

Esar's Comic Dictionary

Support your local medical examiner—die strangely.

Source unknown

Watts, Alan

Life and death are not two opposed forces; they are simply two ways of looking at the same force, for the movement of change is as much the builder as the destroyer.

The Wisdom of Insecurity
Chapter III (p. 41)

Webster, John

I know death hath ten thousand several doors
For men to make their exits.

The Duchess of Amalfi

I'M AFRAID HE'S CURED HIMSELF !

. . . we all labour against our owne cure,
for death is the cure of all diseases.
Sir Thomas Browne – (See p. 58)

DENTAL

Hood, Thomas
Of all our pains, since man was curst,
I mean of body, not the mental,
To name the worst, among the worst,
The dental sure is transcendental;
Some bit of masticating bone,
That ought to help to clear a shelf:
But lets its proper work alone,
And only seems to gnaw itself.

The Poetical Works of Thomas Hood
Volume 1
A True Story
L. 1–8

DENTIST

Bell, H.T.M.
Though many dismal years I've been
To dull old Care apprenticed,
Of smaller woes the worst I've seen
Is—waiting for the dentist!

Source unknown
Waiting for the Dentist

Benjamin, Arthur
We all basically go back to being a child when we're in a dentist's chair.

Newsweek
A Free Bike with Your Braces (p. 82)
May 5, 1986

Bierce Ambrose
Dentist, *n.* A prestidigitator who puts metal into your mouth, pulls coins out of your pocket.

The Enlarged Devil's Dictionary

Chamfort, Nicolas
Our gratitude to most benefactors is the same as our feeling for dentists who have pulled our teeth. We acknowledge the good they have done and the evil from which they have delivered us, but we remember the pain they occasioned and do not love them very much.

Maximes et Pensées

Davies, Robertson
In odd corners of the world strange dentists still lurk; an Irish friend of mine told me recently of visiting a dentist on the West Coast of Ireland who had no running water, and bade his patients spit into a potted fern which was conveniently placed by the chair . . .

The Table Talk of Samuel Marchbanks (p. 178)

Egerton-Warburton, R.E.
In childhood who my first array
Of teeth pluck'd tenderly away,
For teeth like dogs have each their day?
My Dentist.

Source unknown
My Dentist

Editor of the Louisville Journal
A locofoco editor at Brooklyn has quit the business and turned dentist. The poor starveling is unable to find employment for his own teeth except by pulling out those in public.

In George Denison Prentice
Prenticeana (p. 123)

A dentist at work in his vocation always looks down in the mouth.

In George Denison Prentice
Prenticeana (p. 300)

Several young ladies in New Orleans are studying dentistry. We suspect their object is to get near the gentlemen's lips.

In George Denison Prentice
Prenticeana (p. 306)

Fillery, Frank
Dentists' precept: The tooth, the holed tooth, and nothing but the tooth.

Quote, The Weekly Digest
November 12, 1967 (p. 397)

Flaubert, Gustave
Dentists. All untruthful. They us steel balm: are said to be also chiropodists. Pretend to be surgeons, just as opticians pretend to be physicists.

Dictionary of Accepted Ideas

Friedman, Shelby
And there's that tender ballad with a bite: *Since Grandpa Misplaced His False Teeth, He's Been Dentally Retarded!*

Quote, The Weekly Digest
February 25, 1968 (p. 157)

Hale, Susan
He got into my mouth along with a pickaxe and telescope, battering-ram and other instruments, and drove a lawn-cutting machine up and down my jaws for a couple of hours. When he came out he said he meant wonderful improvements, and it seems I'm going to have a bridge and

mill-wheel and summit and crown of gold, and harps, and Lord knows
what.

In Caroline P. Atkinson (Editor)
Letters of Susan Hale
Chapter X
To Mrs. William G. Weld
September 19, 1897 (p. 327)

Hayes, Heather
Epitaph for a dentist: Here lies doctor so an' so filling his last cavity.

Quote, The Weekly Digest
February 25, 1968 (p. 157)

Hood, Thomas
. . . a dentist and the wheel
Of Fortune are a kindred cast,
For after all is drawn, you feel
It's paying for a blank at last: . . .

The Poetical Works of Thomas Hood
Volume 1
A True Story
L. 43–46

Karch, Carroll S.
Today's new drugs, equipment
All ease the dentist's drill
The really painful jab you'll find
Comes from the dentist's bill.

Quote, The Weekly Digest
October 27, 1968 (p. 336)

Kraus, Jack
DENTIST: A member of the Ivory League.

Quote, The Weekly Digest
January 15, 1967 (p. 57)

Lower, Lennie
I've often wondered how people become dentists. Probably some sadistic
urge due to ill-treatment in early youth.

In Cyril Pearl
The Best of Lennie Lower
Charge Your Hypodermics! (p. 211)

Morley, Christopher
The only previous time he had taken gas was in a dentist's office in the Flatiron Building. Whenever he visited that dentist he was always thrilled by the view from the chair, which included the ornate balconies of the old Madison Square Garden and the silhouette of Diana tiptoe in the sky.

Human Being
Pathology (p. 205)

Nash, Ogden
Some tortures are physical and some are mental.
But one that's both is dental.

The Reader's Digest
Have You A Pash for Ogden Nash (p. 10)
July 1952

Perelman, S.J.
I had always thought of dentists as of the phlegmatic type—square-jawed sadists in white aprons who found release in trying out new kinds of burs on my shaky little incisors.

Crazy Like a Fox
Nothing but the Tooth (p. 69)

For years I have let dentists ride roughshod over my teeth; I have been sawed, hacked, chopped, whittled, bewitched, bewildered, tattooed, and signed on again; but this is cuspid's last stand.

Crazy Like a Fox
Nothing but the Tooth (p. 72)

Redford, Sophie E.
There is a time, there is a place,
When he looks down into my face—
And now my senses madly thrill:
He "fills" a place none else can fill!

Without him that dull, aching void
But mocks at me when I'm annoyed
Beyond endurance, by the tricks
of Demons only he can "fix"!

'Tis then he means to me so much!
Vexations vanish by his touch!
And though my list of friends is full,
'Tis he, alone, who has the "pull"!

They say this fellow has much nerve,
That his strong arm must never swerve,
His surplus nerve I think I see
Replenish as he "unnerves" me!

So I would build an arch prodigious
To him who builds the little "bridges",
And fashion then a laurel wreath
To crown the man who "crowns our teeth"!

Cartoon Magazine
To the D.D.S. (p. 829)
Volume 18, Number 6, December 1920

Smith, H. Allen

Barbers were the original dentists, and barbers have traditionally been gabby. So it is that dentists, by retrogression, are usually quite articulate while they work. But there is an important difference between the garrulous dentist and the verbose barber: a customer in the barber's chair can answer back.

Quote, The Weekly Digest
February 19, 1967 (p. 144)

Twain, Mark

Most cursed of all are the dentists who make too many parenthetical remarks—dentists who secure your instant and breathless interest in a tooth by taking a grip on it, and then stand there and drawl through a tedious anecdote before they give the dreaded jerk. Parentheses in literature and dentistry are in bad taste.

A Tramp Abroad
Volume I
Appendix D
The Awful German Language (p. 274)

Some people who can skirt precipices without a tremor have a strong dread of the dentist's chair . . .

Europe and Elsewhere
Down the Rhône (p. 161)

All dentists talk while they work. They have inherited this from their professional ancestors, the barbers.

Europe and Elsewhere
Down the Rhône (p. 162)

Unknown
A dentist named Archibald Moss
Fell in love with the dainty Miss Ross,
   But he held in abhorrence
   Her Christian name, Florence,
So he renamed her his Dental Floss.

In William S. Barring-Gould
The Lure of the Limerick (p. 102)

[Dentist] A man with more pull than a politician.

Esar's Comic Dictionary

[Dentist] A man who always tries not to get on your nerves.

Esar's Comic Dictionary

Members of the seventh grade class (in Louisville) were instructed to write an essay on what they wanted to be when they grew older. One youngster wrote: "I want to be a dentist, like my father because I figure by the time I grow up, he will have all of his equipment paid for."

Quote, The Weekly Digest
October 27, 1968 (p. 334)

. . . Drill, fill and bill.

Newsweek
A Free Bike with Your Braces (p. 82)
May 5, 1986

Watt, William
Her snow-white teeth are, not a little, tinged with the jet;
To the dentist she must go,
And repair the upper row,
Then haply she may run a chance of marriage yet.

Source unknown
Miss Harriot Lucy Brown

Waugh, Evelyn
All this fuss about sleeping together. For physical pleasure I'd sooner go to my dentist any day.

Vile Bodies
Chapter VI (p. 122)

Wells, H.G
. . . . he had one peculiar weakness; he had faced death in many forms but he had never faced a dentist. The thought of dentists gave him just the same sick horror as the thought of invasion.

Bealby
Part VIII
How Bealby Explained (p. 264)

Wilde, Oscar
It is very vulgar to talk like a dentist when one isn't a dentist. It produces a false impression.

The Importance of Being Earnest
Act I

Woolf, Virginia
. . . when we have a tooth out and come to the surface in the dentist's arm-chair and confuse his "Rinse the mouth—rinse the mouth" with the greeting of the Diety stooping from the floor of Heaven to welcome us . . .

The Moment
On Being Ill (p. 9)

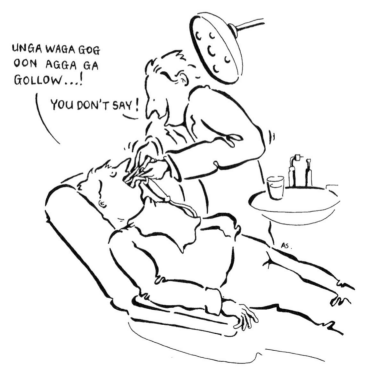

. . .But there is an important difference between the garrulous dentist and the verbose barber: a customer in the barber's chair can answer back.
H. Allen Smith – (See p. 74)

DERMATOLOGIST

Armour, Richard
The obstetrician, roused from bed,
  Gets up all cold and shivery
And has to drive at breakneck speed
  To make a quick delivery.
The surgeon leaves his food untouched
  (The call is full of urgency)
And hastens to the hospital
  To meet a dire emergency.
And both of them, when torn from sleep
  Or when a meal is missed,
Ask, "Why did I not choose to be
  A dermatologist?"

The Medical Muse
When Duty Calls

Frank, Julia Bess
I wish the dermatologist
Were less a firm apologist
For all the terminology
That's used in dermatology.

The New England Journal of Medicine
Dermatology (p. 660)
Volume 297, Number 12, 1977

McLaughlin, Mignon
Psychiatrists are terrible ads for themselves, like a dermatologist with acne.

The Neurotic's Notebook (p. 73)

Unknown
[Dermatologist] A doctor who often makes rash diagnoses.

Esar's Comic Dictionary

[Dermatologist] A specialist who usually starts from scratch.

Esar's Comic Dictionary

Principles of a dermatologist:
If it's wet, dry it.
If it's dry, wet it.
If neither of these works, use steroids.
If steroids don't work, do a biopsy.

Source unknown

Dermatology is the only specialty in medicine where there are 200 diseases and only three types of cream to treat them.

Source unknown

DIAGNOSIS

Bierce, Ambrose
DIAGNOSIS, *n*. The physician's art of determining the condition of the patient's purse, in order to find out how sick to make him.

The Enlarged Devil's Dictionary

DIAGNOSIS, *n*. A physician's forecast of disease by the patient's pulse and purse.

The Enlarged Devil's Dictionary

Eisenschiml, Otto
The best physician must fail if his treatment is based on a wrong diagnosis.

The Art of Worldly Wisdom
Part Ten (p. 119)

Field, Eugene
Upon an average, twice a week,
When anguish clouds my brow,
My good physician and friend I seek
To know "what ails me now".
He taps me on my back and chest,
And scans my tongue for bile,
And lays an ear against my breast
And listens there awhile;
Then is he ready to admit
That all he can observe
Is something wrong inside, to wit:
My pneumosgastric nerve!

The Poems of Eugene Field
The Pneumosgastric Nerve

Foster, Nellis B.
It is an art to secure a complete and accurate history of a patient's sickness. While subject to the laws of logic, securing the history of the

patient is much the same as cross-examination of a witness by an acute lawyer. And the physical examination is wholly an art. The accuracy of the observations made depend entirely upon the technical proficiency of the physician. Knowledge of the significance of physical signs alone is useless unless it be combined with the technical expertness to detect these signs. There is a close analogy between clinical medicine and music. One may know harmony, counterpoint, and all that makes up the science of music, but unless by dint of practice he has mastered technique there will be no music. Technique in music produces beauty of tone, in medicine it secures accuracy of data. There are many sources of error in diagnosis, errors in data, but the commonest of all are errors of technique.

Diagnosis is then a science and an art; a science in the method of using facts secured, an art largely in the mode of collecting facts.

The Examination of Patients

Hoffmann, Friedrich
Who identifies well, treats well; hence the diagnosis of disease is in the highest degree necessary for a physician.

Fundamenta Medicianae
Semiotics
Chapter I, 2 (p. 83)

When investigating the nature of disease, we should attend to all the signs and symptoms. They should not be considered in isolation but rather in combination with each other.

Fundamenta Medicianae
Semiotics
Chapter I, 6 (p. 83)

Holmes, Oliver Wendell
Young doctors are particularly strong on what they call *diagnosis*—an excellent branch of the healing art, full of satisfaction to the curious practitioner who likes to give the right Latin name to one's complaints; not quite so satisfactory to the patient, as it is not so very much pleasanter to be bitten by a dog with a collar round his neck telling you that he is called *Snap* or *Teaser*, than by a dog without a collar. Sometimes, in fact, one would a little rather not know the exact name of his complaint, as if he does he is pretty sure to look it out in a medical dictionary, and then if he reads, "This terrible disease is attended with vast suffering and is inevitably mortal," or any such statement, it is apt to affect him unpleasantly.

The Poet at the Breakfast Table

Hutchison, Sir Robert
It may be granted freely that a bad diagnosis due to an error of judgment is more excusable than one attributable to want of knowledge or even

to faulty observation. The ghosts of dead patients which at the midnight hour haunt the bedside of every doctor who has been some years in practice will not upbraid him with such questions as "Why did you not know that a ball-valve gall stone may produce symptoms like those of malaria?" or still less, "Why did you not attach more importance to the rapidity of my pulse, and less to the signs in my abdomen?" No; the unescapable questions they will put to him will be such as these: "Why did you not examine my fundi for optic neuritis?" or "Why did you not put a finger in my rectum?"

British Medical Journal
The Principles of Diagnosis
1928

Kraus, Karl
One of the most widespread diseases is diagnosis.

Half-Truths & One-and-a-Half Truths (p. 77)

Latham, Peter Mere
The diagnosis of disease is often easy, often difficult, and often impossible.

In William B. Bean
Aphorisms from Latham (p. 56)

Ordinary diseases will sometimes occur under extraordinary circumstances, or in unusual situations; and then we are apt to be thrown out in our diagnosis, as the pilot is in his course upon any unexpected alteration of lights and signals on the coast. He makes false points, and so do we.

In William B. Bean
Aphorisms from Latham (p. 57)

You may listen to the chest for ever and be no wiser, if you do not previously know what it is you are to hear. You may beat the chest for ever, and all in vain, unless you know what it is that is capable of rendering it now dull and now resonant.

In William B. Bean
Aphorisms from Latham (p. 58)

Osler, Sir William
If necessary, be cruel; use the knife and the cautery to cure the intumescence and moral necrosis which you will feel in the posterior parietal region, in Gall and Spurzheim's center of self-esteem, where you will find a sore spot after you have made a mistake in diagnosis.

In Harvey Cushing
The Life of Sir William Osler
Volume I (p. 329)

Unknown
[Diagnosis] The preface to an autopsy.

Esar's Comic Dictionary

A true diagnosis of a case contains in itself the therapy.

In Ernst Lehrs
Man or Matter (p. 30)

One finger in the throat and one in the rectum
makes a good diagnostician.
Sir William Osler – (See opposite)

DIAGNOSTICIAN

Osler, Sir William
One finger in the throat and one in the rectum makes a good diagnostician.

Aphorisms from his Bedside Teachings (p. 104)

Hutchison, Sir Robert
Don'ts for Diagnosticians
1. Don't be too clever.
2. Don't diagnose rarities.
3. Don't be in a hurry.
4. Don't be faddy.
5. Don't mistake a label for a diagnosis.
6. Don't diagnose two diseases simultaneously in the same patient.
7. Don't be too cock-sure.
8. Don't be biased.
9. Don't hesitate to revise your diagnosis from time to time in a chronic case.

British Medical Journal
The Principles of Diagnosis
1928

DIAPHRAGM

Bierce, Ambrose

DIAPHRAGM, *n*. A muscular partition separating disorders of the chest from disorders of the bowels.

The Enlarged Devil's Dictionary

DIET

Arnoldus

A wise Physition will not give Physicke, but upon necessity, and first try medicinall diet, before hee proceede to medicinall cure.

In Robert Burton
The Anatomy of Melancholy
Volume II
Part 2, Section 1, memb. 4, subs. 1 (p. 75)

Bacon, Francis

Beware of sudden changes in any great point of diet, and, if necessity inforce it, fit the rest to it.

Essays, Advancement of Learning, New Atlantis, and Other Pieces
The Essayes or Counsels, Civil and Morall
XXX, Of Regiment of Health (p. 93)

Davis, Adelle

When the blood sugar is extremely low, the resulting irritability, nervous tension, and mental depression are such that a person can easily go berserk . . . Add a few guns, gas jets, or razor blades, and you have the stuff murders and suicides are made of. The American diet has become dangerous in many more ways than one.

Let's Eat Right to Keep Fit
Chapter 2 (p. 19)

Editor of the Louisville Journal

What some call health, if purchased by perpetual anxiety about diet, isn't much better than tedious disease.

In George Denison Prentice
Prenticeana (p. 302)

Grant, Claud

Diets are for people who are thick and tired of it.

Quote, The Weekly Digest
June 2, 1968 (p. 437)

Propp, Fred Jr.
Diet: Slowing down to make a curve.

Quote, The Weekly Digest
July 23, 1967 (p. 77)

Proverb, Chinese
He that takes medicine and neglects to diet wastes the skill of his doctor.

Source unknown

Simmons, Charles
He who eats of but one dish, never wants a physician.

Laconic Manual and Brief Remarker (p. 234)

When I behold a fashionable table, set out in all its magnificence, I fancy that I see gouts and dropsies, fevers and lethargies, with innumerable distempers, lying in ambuscade among the dishes. Nature delights in the most plain and simple diet. Every animal, but man, keeps to one dish. Herbs are the food of this species, fish of that, and flesh of a third. Man falls upon everything that comes his way; not the smallest fruit or excrescence of the earth, scarce a berry or mushroom can escape him.

Laconic Manual and Brief Remarker (p. 486)

Unknown
Spare diet and no trouble keep a man in good health.

Source unknown

[Diet] A form of wishful shrinking.

Esar's Comic Dictionary

There was a young lady who tried
A diet of apples and died.
    The unfortunate miss
    Really perished of this:
Too much cider inside her inside.

In Louis Untermeyer
Lots of Limericks (p. 136)

DIGESTION

Chesterton, Gilbert Keith
Digestion exists for health, and health exists for life, and life exists for the love of music or beautiful things.

Generally Speaking
On Misunderstanding

DISCHARGE

Flaubert, Gustave

Discharge. Rejoice when it leaves any affected part, and express astonishment that the human body can contain so much matter.

Dictionary of Accepted Ideas

DISEASE

Bacon, Francis
. . . diseases of the body may have appropriate exercises; bowling is good for the stone and veins, shooting for the lungs and breast, gentle walking for the stomach, riding for the head and the like . . .

Bacon's Essays
Of Studies (p. 129)

. . . cure the disease and kill the patient.

Bacon's Essays
Of Friendship (p. 74)

Barach, Alvin
Remember to cure the patient as well as the disease.

Recalled on his death
December 15, 1977

Baring, Maurice
Pale disease
Shall linger by thy side, and thou shalt know
Eternal autumn to thy day of death.

The Black Prince and Other Poems
The Black Prince and the Astrologer (p. 59)

Belloc, Hilaire
Physicians of the Utmost Fame
Were called at once; but when they came
They answered, as they took their Fees,
'There is no Cure for this Disease.'

Cautionary Tales for Children
Henry King (pp. 18–19)

Bierce, Ambrose
DISEASE, *n*. Nature's endowment of medical schools. A liberal provision for the maintenance of undertakers. A means of applying the worthy

grave-worm with meat that is not too dry and tough for tunneling and stopping.

The Enlarged Devil's Dictionary

Browne, Sir Thomas
. . . medical Predictions fail not, as sometimes in acute Diseases, and wherein 'tis as dangerous to be sentenced by a Physician as a Judge.

The Works of Sir Thomas Browne
Volume Three
A Letter to a Friend (p. 370)

Some will allow no Disease to be new, others think that many old ones are ceased; and that such which are esteemed new, will have but their time.

The Works of Sir Thomas Browne
Volume One
A Letter to a Friend (pp. 172–3)

Byron, Lord George Gordon
When slow Disease, with all her host of pains,
Chills the warm tide which flows along the veins;
When Health, affrighted, spreads her rosy wing,
And flies with every changing gale of spring; . . .

The Poetical Works of Lord Byron
Childish Recollections (p. 550)

Carlyle, Thomas
Self-contemplation . . . is infallibly the symptom of disease . . .

Characteristics (p. 17)

Chekhov, Anton
People love talking of their diseases, although they are the most uninteresting things in their lives.

Note-Book of Anton Chekhov (p. 28)

Churchill, Winston
The discoveries of healing science must be the inheritance of all. That is clear. Disease must be attacked, whether it occurs in the poorest or the richest man or woman, simply on the ground that it is the enemy; and it must be attacked just in the same way as the fire brigade will give its full assistance to the humblest cottage as readily as to the most important mansion.

In F.B. Czarnomski
The Wisdom of Winston Churchill (p. 171)
Speech
Royal College of Physicians, London, March 2, 1944

Deuteronomy 28:22
The Lord shall smite thee with a consumption, and with a fever, and with an inflammation, and with an extreme burning . . .

The Bible

Dryden, John
No Sickness known before, no slow Disease,
To soften Grief by Just Degrees.

The Poems of John Dryden
Volume I
Threnodia Augustalis
Stanza 1 (p. 442)

Eddy, Mary Baker
. . . disease can carry its ill-effects no farther than mortal mind maps out the way.

Science and Health, with Key to the Scriptures (p. 176)

Fielding, Henry
Every physician almost hath his favorite disease.

Tom Jones
Book II, Chapter 9 (p. 33)

Fuller, Thomas
Diseases are the price of ill pleasures.

Gnomologia
Number 1297

Harrison, Jane
If I think of Death at all it is merely as a negation of life, a close, a last and necessary chord. What I dread is disease, that is, bad, disordered life, not Death, and disease, so far, I have escaped.

Reminiscences of a Student's Life
Conclusion

Heber, Reginald
Each season has its own disease,
Its peril every hour.

The Poetical Works of Bishop Heber
At a Funeral

Heller, Joseph
Hungry Joe collected lists of fatal diseases and arranged them in alphabetical order so that he could put his finger without delay on any one he wanted to worry about.

Catch-22
Chapter 17 (p. 170)

James, Henry
. . . even medical families cannot escape the more insidious forms of disease . . .

Washington Square
Chapter I (p. 4)

Jerome, Jerome K.
I remember going to the British Museum one day to read up the treatment for some slight ailment of which I had a touch. I got down the book, and read all I came to read; and then, in an unthinking moment, I idly turned the leaves, and began to indolently study diseases. Bright's disease, I was relieved to find, I had only in a modified form, and so far as that was concerned, I might live for years. Cholera I had, with severe complications: and diphtheria I seemed to have been born with. I plodded conscientiously through the twenty-six letters, and the only malady I could conclude I had not got was housemaid's knee.

Three Men in a Boat
Chapter 1 (p. 3)

Jhabvala, Ruth Prawer
"Doctors don't know a thing. These diseases that people get in India, they're not physical, they're purely psychic. We only get them because we try to resist India—because we shut ourselves up in our little Western egos and don't want to give ourselves."

Travelers (p. 166)

Job 2:7
So went Satan forth from the presence of the Lord, and smote Job with sore boils from the sole of his foot unto his crown.

The Bible

Latham, Peter Mere
Disease is a series of new and extraordinary actions. Each link in the series is essential to the integrity of the whole. Let one link be fairly broken, and this integrity is spoiled; and there is an end of the disease; and then the constitution is left to resume its old and accustomed actions, which are the actions of health.

In William B. Bean
Aphorisms from Latham (p. 71)

Mather, Cotton
Ingluvies omnium morborum mortisque causa.
[Gluttony is the cause of all diseases and of death.]

The Angel of Bethesda
Capsila II, Appendix (p. 15)

Montaigne, Michel de
For a desperate disease a desperate cure.

Essays
Book the Second
Chapter 3 (p. 167)

Nightingale, Florence
But when you have done away with all that pain and suffering, which in patients are the symptoms not of their disease, but of the absence of one or all of the above-mentioned essentials to the success of Nature's reparative processes, we shall then know what are the symptoms of and the sufferings inseparable from, the disease.

Notes on Nursing
Notes on Nursing (p. 9)

Persius
. . . check the ailment before it's got to you . . .

The Satires of Persius
Satire Three, L. 67

Plato
". . . and to require the help of medicine, not when a wound has to be cured, or on occasion of an epidemic, but just because, by indolence and habit of life such as we have been describing, men fill themselves with waters and winds, as if their bodies were a marsh, compelling the ingenious sons of Asclepius to find more names for diseases, such as flatulence and catarrh; is not this, too, a disgrace?"

"Yes," he said, "they do certainly give very strange and new-fangled names to diseases."

"Yes," I said, "and I do not believe there were any such diseases in the days of Asclepius."

The Republic
Book III [405] (p. 335)

Proverb, Latin
Occultare morbum funestam.
[To hide disease is fatal.]

Source unknown

Ray, John
Diseases are the interests of pleasures.

A Complete Collection of English Proverbs (p. 6)

Rogers, Will

We were primitive people when I was a kid. There were only a mighty few known diseases. Gunshot wounds, broken legs, toothache, fits, and anything that hurt you from the lower end of your neck down was known as a bellyache.

The Autobiography of Will Rogers
Chapter Twelve (p. 151)

Sacks, Oliver

Diseases have a character of their own, but they also partake of our character; we have a character of our own, but we also partake of the world's character . . .

Awakenings
Perspectives (p. 206)

Seneca

. . . a disease also is farther on the road to being cured when it breaks forth from concealment and manifests its power.

Ad Lucilium Epistulae Morales
Volume I
Epistle lvi, Section 10 (p. 379)

Shadwell, Thomas

Physicians tell us, that in every Age
Some one particular Disease does rage,
The Scurvy once, and what you call the Gout,
But Heaven be prais'd their Reign is almost out . . .

The Complete Works of Thomas Shadwell
The Sullen Lovers
Epilogue

Shakespeare, William

Diseases desperate grown
By desperate appliance are relieved,
Or not at all.

Hamlet
Act IV, Scene III, L. 9–11

O, he's a limb, that has but a disease;
Mortal, to cut it off; to cure it easy.

Coriolanus
Act III, Scene I, L. 296–297

I'll sweat and seek about for eases,
And at that time bequeath you my diseases.

Troilus and Cressida
Act V, Scene X, L. 56–57

Shaw, George Bernard

There is at bottom only one genuinely scientific treatment for all diseases, and that is to stimulate the phagocytes.

The Doctor's Dilemma
Act I (p. 28)

Simmons, Charles

The diseases and "evils which flesh is heir to," are all the messengers of God, to rebuke us for our sins, and ought so to be regarded.

Laconic Manual and Brief Remarker (p. 148)

Thurber, James

If you don't pay no mind to diseases, they will go away.

The Thurber Carnival
Recollections of the Gas Buggy (p. 36)

Twain, Mark

The human being is a machine. An automatic machine. It is composed of thousands of complex and delicate mechanisms, which perform their functions harmoniously and perfectly, in accordance with laws devised for their governance, and over which the man himself has no authority, no mastership, no control. For each one of these thousands of mechanisms the Creator has planned an enemy, whose office is to harass it, pester it, persecute it, damage it, afflict it with pains, and miseries, and ultimate destruction. Not one has been overlooked.

From cradle to grave these enemies are always at work; they know no rest, night or day. They are an army: an organized army; a besieging army; an assaulting army; an army that is alert, watchful, eager, merciless; an army that never relents, never grants a truce. It moves by squad, by company, by battalion, by regiment, by brigade, by division, by army corps; upon occasion it masses its parts and moves upon mankind with its whole strength. It is the Creator's Grand Army, and he is the Commander-in-Chief. Along its battlefront its grisly banners wave their legends in the face of the sun: Disaster, Disease, and the rest.

Disease! That is the force, the diligent force, the devastating force! It attacks the infant the moment it is born; it furnishes it one malady after another: croup, measles, mumps, bowel troubles, teething pains, scarlet fever, and other childhood specialties. It chases the child into youth and furnishes it some specialties for that time of life. It chases the youth into maturity, maturity into age, and age into the grave.

Letters from the Earth
Letter VI (pp. 28–9)

Unamuno, Miguel de
"There are no diseases, but only persons who are diseased," some doctors say, and I say that there are no opinions, but only opining persons.

Essays and Soliloquies (pp. 156–7)

Unknown
Love passes, but syphilis remains.

In Evan Esar
20,000 Quips and Quotes

Doctors today say that fatal diseases are the worst.

In Alexander Abingdon
Bigger & Better Boners (p. 71)

An imaginary ailment is worse than a disease.

Source unknown

Viereck, George Sylvester
Eldridge, Paul
All diseases are curable, provided the patient lives long enough to overcome the initial cause of the complaint.

My First Two Thousand Years
Chapter VII (p. 56)

Young, Arthur
Catch the disease, that we may show our skill in curing it!

The Adventures of Emmera
Volume II
Letter 26 (p. 115)

Young, Edward
Polite disease make some ideots *vain*,
Which, if unfortunately well, they feign.

Love of Fame
Satire I, L. 95–96 (p. 9)

DISINFECTANT

Osler, Sir William
Soap and water and common sense are the best disinfectants.

In Evan Esar
20,000 Quips and Quotes

DISSECTION

Flaubert, Gustave
Dissection. An outrage upon the majesty of death.

Dictionary of Accepted Ideas

DOCTOR

Albricht, Fuller
As with eggs, there is no such thing as a poor doctor, doctors are either good or bad.

Russell L. Cecil and Robert F. Leob
Textbook of Medicine
9th edition
Diseases of the Ductless Glands
Introduction

Amiel, Henri-Frédéric
. . . the ideal doctor would be a man endowed with profound knowledge of life and of the soul, intuitively divining any suffering or disorder of whatever kind, and restoring peace by his mere presence.

Amiel's Journal
August 20, 1873 (p. 263)

Armour, Richard
Oh, stay away from the doctor my friend,
And suffer in silence your ills,
You'll save all that waiting in waiting rooms
And save on your doctor bills.
If you have a pain and it's getting worse,
You can tell your neighbor, it's true,
As long as your neighbor is not an M.D.
Or likely to rat on you.

Quote, The Weekly Digest
Taking No Chances (p. 278)
October 1, 1967

Look up noses,
   Look down throats,
Look up nostrums,
   Jot down notes,

Look up rectums,
  Look down ears,
Look up patients
  In arrears . . .

The Medical Muse
The Doctor's Life

Arnold, Matthew

Nor bring, to see me cease to live,
Some doctor full of phrase and fame,
To shake his sapient head, and give
The ill he cannot cure a name.

Poems by Matthew Arnold
Volume Two
A Wish

Bacon, Francis

We know diseases of stoppings and suffocations are the most dangerous
in the body . . . you may take sarza to open the liver, steel to open the
spleen, flower of sulphur for the lungs, castoreum for the brain . . .

Bacon's Essays
Of Friendship (p. 70)

Balzac, Honoré de

Like the rod of Moses, the doctor's orders make and unmake generations.

The Physiology of Marriage
Part Three
Chapter 25, Section 5 (pp. 302)

All doctors know what great influence women have on their reputation;
thus we meet with few doctors who do not study to please the ladies.

The Physiology of Marriage
Part Three
Chapter 25, Section 5 (pp. 302–3)

Bierce, Ambrose

DOCTOR, *n.* A gentleman who thrives upon disease and dies of health.

The Enlarged Devil's Dictionary

Booth, John

A single doctor like a sculler plies,
And all his art and all his physics tries;
But two physicians, like a pair of oars,
Conducts you soonest to the Stygian shores.

Epigrams Ancient and Modern (p. 144)

Bradford, Maynerd

A doctor had just hired a new secretary. Having trouble with the doctor's notes on an emergency case which read, "Shot in the lumbar region," the poor girl was flustered and at wits end. At last she thought she had it figured out and brightened up as she typed up the record, "Wounded in the woods."

Quote, The Weekly Digest
September 8, 1968 (p. 195)

Brain, W. Russell

The doctor occupies a seat in the front row of the stalls of the human drama, and is constantly watching, and even intervening in, the tragedies, comedies and tragi-comedies which form the raw material of the literary art.

Foreword to Robert Coope
The Quiet Art

Butler, Samuel

We in England never shrink from telling our doctor what is the matter with us merely through the fear that he will hurt us. We let him do his worst upon us, and stand it without a murmur, because we are not scouted for being ill, and because we know that the doctor is doing his best to cure us, and that he can judge of our case better than we can . . .

Erewhon
Chapter X (p. 100)

Cairns, Sir Hugh

How does one become a good doctor? When one doctor says of another, "He is a good doctor," the words have a particular meaning. You will hear the expression used not only about some general practitioners, but also about some specialists. As I understand it a good doctor is one who is shrewd in diagnosis and wise treatment; but, more than that, he is a person who never spares himself in the interest of his patients; and in addition he is a man who studies the patient not only as a case but also as an individual.

The Lancet
The Student's Objective (p. 665)
Volume 257, October 8, 1949

Caldwell, George W.

Who fights disease, and death, and fear,
Through all our life to keep us here?
And when we draw our final breath,
Who is our faithful friend till death—
The doctor!

In Mary Lou McDonough
Poet Physician
The Doctor (p. 137)

Causley, Charles
You must take off your clothes for the doctor
And stand as straight as a pin,
His hand of stone on your white breastbone
Where the bullets all go in.

Collected Poems
Recruiting Drive

Chaplin, Albin
A doctor said sadly, "Alas!
From the data that I can amass,
  What causes male droop—
  And I have all the poop—
Is the feminine excess of ass."

In G. Legman (Editor)
The New Limerick
Little Romances (p. 1)

Chekhov, Anton
There has been an increase not in the number of nervous diseases and nervous patients, but in the number of doctors able to study those diseases.

Note-Book of Anton Chekhov (pp. 25–6)

The only result he gets is to hate doctors.

Note-Book of Anton Chekhov (p. 44)

A conversation at a conference of doctors. First doctor: "All diseases can be cured by salt." Second doctor, military: "Every disease can be cured by prescribing no salt."

Note-Book of Anton Chekhov (p. 69)

The doctor said to me: "If," says he, "Your constitution holds out, drink to your heart's content."

Note-Book of Anton Chekhov (p. 81)

Only one thought reconciled him to the doctor: just as he had suffered from the doctor's ignorance, so perhaps some one was suffering from his mistakes.

Note-Book of Anton Chekhov (p. 145)

Christie, Agatha
Doctors can do almost anything nowadays, can't they, unless they kill you first while they're trying to cure you.

Endless Night
Chapter 11 (p. 98)

Cordus, Euricus
Three faces wears the doctor: when first sought
An angel's!—and a god's the cure half wrought:
But when, that cure complete, he seeks his fee,
The Devil then looks less terrible than he.

Annals of Medical History
The Three Characters of a Physician (p. 53)
Volume I, 1917

Corrigan, Sir Dominic J.
The trouble with doctors is not that they don't know enough, but that they don't see enough.

In Robert Coope
The Quiet Art (p. 107)

Crichton-Browne, Sir James
The young doctor of to-day is undoubtedly far better educated and more skilled and scientific than the doctor of a century ago, but with his motor car and his five hundred panel patients his practice is apt to become a little mechanical, except for interesting cases, and to lack the intimate personal knowledge and sympathy of the old leisurely family doctor in his gig. Hustle is the danger of the day.

The Doctor's After Thoughts (p. 172)

da Costa, J. Chalmers
No doctor should be jealous of another doctor. He should be like the seasick man—not desirous of other people's things and only anxious to keep what he has.

The Trials and Triumphs of the Surgeon
Stepping Stones and Stumbling Blocks
Part III (p. 230)

da Vinci, Leonardo
Every man desires to acquire wealth in order that he may give it to the doctors, the destroyers of life; therefore they ought to be rich.

Leonardo da Vinci's Notebooks (p. 65)

de La Fontaine, Jean
Dr It-can't-be-helped and Dr It-can
Met at the bedside of an ailing man.

La Fontaine: Selected Fables
Book V
The Doctors

Dickens, Charles
There might be medical doctors at the present hour, a-picking up their guineas where a honest tradesman don't pick up his fardens—fardens!

no, nor yet his quarter—a-banking away like smoke at Tellson's, and a-cocking their medical eyes at that tradesman on the sly . . .

A Tale of Two Cities
Book 3, Chapter IX (p. 326)

Dryden, John
So liv'd our Sires, 'ere Doctors learn'd to kill,
And multiply'd with theirs, the Weekly Bill: . . .

The Poems of John Dryden
Volume IV
To John Dryden, of Chesterton
L. 71–72

Dubos, René
In the words of a wise physician, it is part of the doctor's function to make it possible for his patients to go on doing the pleasant things that are bad for them—smoking too much, eating too much, drinking too much—without killing themselves any sooner than is necessary.

Mirage of Health
Chapter VI (pp. 179–80)

Dubos, René
Escande, Jean-Paul
Doctors have a whole arsenal of useful weapons at their disposal. We mustn't ask them to use only the most spectacular ones.

Quest: Reflections on Medicine, Science, and Humanity
Chapter I (p. 15)

Dunne, Finley Peter
"I think," said Mr. Dooley, "that if th' Christyan Scientists had some science and th' doctors more Christyanity, it wudden't make anny diff'rence which ye called in—if ye had a good nurse."

Mr. Dooley's Opinions
Christian Science (p. 9)

. . . th' dock who shortens his prescriptions lenthens th' lives iv his patients.

Mr. Dooley: On Making a Will and Other Necessary Evils
Going to See the Doctor (p. 122)

A patient in th' hands iv a doctor is like a hero in th' hands iv a story writer. He's goin' to suffer a good dale, but he's goin' to come out all right in th' end.

Mr. Dooley: On Making A Will and Other Necessary Evils
Going to See the Doctor (p. 127)

Editor of the Louisville Journal
The doctors ought surely to be able to escape calumny. It is held that no men *living* should speak ill of them, and the dead *can't*.

In George Denison Prentice
Prenticeana (p. 101)

The doctor is not unfrequently Death's pilot fish.

In George Denison Prentice
Prenticeana (p. 272)

Emerson, Ralph Waldo
On Wachusett, I sprained my foot. It was slow to heal, & I went to the doctors. Dr. H. Bigelow said, "a splint & absolute rest;" Dr. Russell said, "rest yes; but a splint, no." Dr. Bartlett said, "neither splint nor rest, but go & walk." Dr. Russell said, "pour water on the foot, but it must be warm." Dr. Jackson said, "stand in a trout brook all day."

Journal
August 16–19, 1859 (p. 223)

Good is a good doctor, but Bad is sometimes better.

Complete Works
Volume 6
Considerations by the Way

Epigram
Si tibi deficiant Medici, Medici tibi fiant
Haec Tria, Mens laeta, Requies, Moderata Diaeta.
[If doctors fail you, let these doctors work for you,
These three, a happy spirit, rest, and a moderate diet.]

The Angel of Bethesda
Capsila II
Appendix (p. 16)

Esar, Evan
A doctor is a healthy man who can't keep away from sick people.

20,000 Quips and Quotes

When a patient is at death's door, it is the duty of his doctor to pull him through.

20,000 Quips and Quotes

A doctor is the only salesman who always finds his clients in a moment of weakness.

20,000 Quips and Quotes

Doctors should let the well enough alone.

20,000 Quips and Quotes

The man who gets sick calls a doctor, but the man who becomes ill summons a physician.

20,000 Quips and Quotes

Faulkner, William
What do doctors know? They make their livings advising people to do whatever they are not doing at the time, which is the extent of anyone's knowledge of the degenerate ape.

The Sound and the Fury
April Sixth 1928 (p. 200)

Fillery, Frank
While still a boy, he thought that he
Would like to call himself M.D.
Though, then, he really didn't know
That M.D. simply means "more dough."

Quote, The Weekly Digest
February 12, 1967

Flaubert, Gustave
DOCTOR: Always preceded by 'the good.' Among men, in familiar conversation, 'Oh! balls, doctor!' Is a wizard when he enjoys your confidence, a jackass when you're no longer on terms. All are materialists: 'You can't probe for faith with a scalpel.'

Dictionary of Accepted Ideas

Franklin, Benjamin
Beware of the young doctor and the old barber.

Poor Richard
May 1733

Don't go to the doctor with every distemper, nor to the lawyer with every quarrel, nor to the pot for every thirst.

Poor Richard
November 1737

Free, Spencer Michael
Oh, where is the doctor who treated our ills
With Jalap and Rhubard and Senna and squills
A much-whispered man—with the shaggy eyebrows
Who didn't know all of the "whys" and the "hows"
But had much common sense—and a heart that was big
He rode on a horse or sometimes in a gig.

In Mary Lou McDonough
Poet Physician
The Old-Time Family Doctor (p. 108)

Fuller, Thomas
The patient is not likely to recover who makes the doctor his heir.

Gnomologia
Number 4368

Gay, John
Is there no hope, the sick man said,
The silent doctor shook his head,
And took his leave, with signs of sorrow,
Despairing of his fee tomorrow.

Fables I
The Sick Man and the Angel (p. 91)

Gould, Donald
Most doctors are not particularly interested in health. By inclination and training they are devoted to the study of disease. It is sick, not healthy people, who crowd their surgeries and out-patient departments and fill their hospital beds, and it is the fact that the population can be relied upon to provide a steady flow of sufferers from faults of the mind and of the flesh that guarantees them a job and an income in harsh times and fair.

The Black and White Medicine Show
Chapter Two (p. 33)

Guest, Edgar A.
The doctor leads a busy life, he wages war with death;
Long hard hours he spends to help the ones who's fighting hard
for breath;
He cannot call his time his own, nor share in others' fun,
His duties claim him through the night when others' work is done.

Collected Verse of Edgar A. Guest
It's a Boy (p. 311)

They said he was a doctor of six or seven months ago,
They gave him a diploma he could frame and proudly show,
And they said: "Go out and practice and just show 'em what you know."

Collected Verse of Edward A. Guest
The Young Doctor (p. 695)

Gunderson, Gunnar
While the patient wants the best and most modern treatment available, he is also badly in need of the old-fashioned friend that a doctor has always personified and which you must continue to be.

Commencement address
Strich School of Medicine
Loyola University
June 7, 1962

Haliburton, Thomas C.
. . . it's my theory that more folks die of the doctor than the disease.

Sam Slick's Wise Saws
Chapter XI (p. 215)

She went into the doctor's shop some medicine for to find,
Sayin, "Have you any medicine wid knock an aul' man blind?"

Oh, the doctor gave her marrow bones and bid her grind them fine,
And dust them in the aul' man's eyes and that would knock him blind.

In Sam Henry
Songs of the People
The Auld Man And The Churnstaff

Helmuth, William Tod
Ye Doctors, ye must *patient* be
To cure your *patients'* ills;
Or *patient patients* will you see
im*patient* of their bills.

Scratches of a Surgeon
The Great Easter Celebration (p. 111)

Herrick, Robert
When the artlesse Doctor sees
No one hope, but of his Fees,
And his skill runs on the lees;
Sweet Spirit comfort me!

The Complete Poems of Robert Herrick
Volume III
His Litanie, to the Holy Spirit (p. 132)

Hewitt, Barnard
And we cut away to suit ourselves! If a shoemaker spoils a scrap of
leather, he has to pay for it. But a doctor can spoil a man with complete
impunity. A mistake is always blamed on the dead patient!

The Doctor in Spite of Himself
Act III (p. 69)

And most conveniently for the medical profession, there exists among
the dead the highest integrity, the greatest discretion in the world. Not
a single one of them has ever lodged a complaint against his doctor.

The Doctor in Spite of Himself
Act III (p. 69)

Himes, Isaac N.
Slowly, slowly, slowly we oxidize!
Become old and rusty,
Fungoid and musty,
Diminish in size;
Reputation decreases, and self-conceit ceases—
Cares fret and wear out facial lines incessantly,
Yet doctors grow old rather pleasantly!

> In Mary Lou McDonough
> *Poet Physician*
> The Doctor's Life (p. 88)

Holmes, Oliver Wendell
—Talk of your science! after all is said
There's nothing like a bare and shiny head;
Age lends the graces that are sure to please;
Folks want their doctors mouldy, like their cheese.

> *The Complete Poetical Works of Oliver Wendell Holmes*
> Rip Van Winkle, M.D.
> Canto Second

Doctors are oxydable products, and the schools must keep furnishing new ones as the old ones turn into oxyds; some of first-rate quality that burn with great light,—some of a lower grade of brilliancy, some honestly, unmistakably, by the grace of God, of moderate gifts, or in simpler phrase, dull.

> *Medical Essays*
> Scholastic and Bedside Teaching (p. 301)

God help all little children in the hands of dosing doctors and howling dervishes!

> *Medical Essays*
> The Medical Profession in Massachusetts (p. 334)

Hood, Thomas
In fact, he did not find M.D.'s
Worth one D—M.

> *The Poetical Works of Thomas Hood*
> Volume 1
> Jack Hall (p. 93)

There once was a Doctor,
(No foe to the proctor,)
A physic concocter,
Whose dose was so pat,
However it acted,

One speech it extracted,—
"Yes, yes," said the doctor,
"I meant it for that!"

The Poetical Works of Thomas Hood
Volume 2
The Doctor (p. 325)

Howard, Sidney

Oh, it's easy to fool a woman. But you can't fool a doctor . . .

Famous American Plays of the 1920s and the 1930s.
They Knew What They Wanted
Act Three (p. 126)

Hubbard, Kin

Mrs. Em Moots wuz taken dangerously ill lately an' a doctor wuz finally
rounded up who promised t' call within a few days.

Abe Martin: Hoss Sense and Nonsense (p. 25)

James, Alice

I suppose one has a greater sense of intellectual degradation after an
interview with a doctor than from any human experience.

The Diary of Alice James
Letter to William James (September 27, 1890) (p. 142)

Kafka, Franz

Certainly doctors are stupid, or rather they are not more stupid than
other people but their pretensions are ridiculous, nevertheless one has to
count with the fact that they become more and more stupid the moment
one is in their hands and what the doctor demands for the moment is
neither very stupid nor impossible.

Letters to Milena (p. 31)

Karmel, Marjorie

Who ever said that doctors are truthful—or even intelligent? You're
getting a lot if they know their profession. Don't ask any more from
them. They're only human after all—which is to say, you can't expect
much.

Thank You, Dr. Lamaze
Chapter 7 (p. 100)

King, Alexander

With a doctor, you can afford to appear slightly hoydenish, just enough
to make him think he's saving you from something worse than death.

Rich Man, Poor Man, Freud and Fruit
Chapter IV (p. 50)

All doctors have a mania for bawdy jokes, and once they have led the way, it is perfectly safe for you to add to their repertoire.

Rich Man, Poor Man, Freud and Fruit
Chapter IV (p. 51)

Kipling, Rudyard

. . . there are only two classes of mankind in the world—doctors and patients . . .

A Book of Words
A Doctor's Work (p. 43)

It was long ago decided that you have no working hours which anybody is bound to respect, and nothing except for your extreme bodily illness will excuse you in its eyes from refusing to help a man, who thinks he may need your help, at any hour of the day or night. Nobody will care whether you are in your bed, or in your bath, or at the theatre. If any one of the children of men has a pain or a hurt in him you will be summoned; and, as you know, what little vitality you may have accumulated in your leisure will be dragged out of you again.

In all time of flood, fire, famine, plague, pestilence, battle, murder, and sudden death it will be required of you that you report for duty at once, and go on duty at once, and that you stay on duty until your strength fails you or your conscience relieves you; whichever may be the longer period . . . Have you heard of any Bill for an eight hours' day for doctors? . . . you belong to the privileged classes . . . May I remind you of some of your privileges? You and Kings are the only people whose explanation the Police will accept if you exceed the legal limit in your car. On presentation of your visiting-card you can pass through the most turbulent crowd unmolested and even with applause. If you fly a yellow flag over a centre of population you can turn it into a desert. If you choose to fly a Red Cross flag over a desert you can turn it into a centre of population towards which, as I have seen, men will crawl on hands and knees. You can forbid any ship to enter any port in the world. If you think it necessary to the success of any operation in which you are interested, you can stop a 20,000-ton liner with mails in mid-ocean till the operation is concluded. You can order whole quarters of a city to be pulled down or burnt up; and you can trust to the armed co-operation of the nearest troops to see that your prescriptions are properly carried out.

A Book of Words
A Doctor's Work (pp. 44–5)

La Bruyère, Jean

Physicians have been attacked for a long time, and yet everyone consults them; neither the sallies of the stage nor of satire diminish their fees; they give dowries to their daughters, have sons magistrates and bishops; and all this is paid for by the very persons who make fun of them. People who are in good health fall ill some day or other, and then they want a man whose trade it is to assure them they shall not die. As long as men are liable to die, and are desirous to live, a physician will be made fun of, but he will be well paid.

The Characters
Of Certain Customs, 65 (p. 279)

Those who are well get sick; they need people whose business it is to assure them they won't die: as long as men go on dying, and love living, the doctor will be made game of and well-paid.

The Characters
Chapter XIV

Quacks are rash, and therefore rarely successful; hence physic and physicians are in vogue; the latter let you die, the former kill you.

The Characters
Of Certain Customs, 67 (p. 279)

Lamont, Thomas W.
When I was young and full of life
I loved the local doctor's wife,
And ate an apple every day
To keep the doctor far away.

My Boyhood in a Parsonage
Chapter V (p. 29)

Martial
Diaulus has been a doctor, he is now an undertaker. He begins to put his patients to bed in his old effective way.

Epigrams
Book I, epigram XXX

Lately was Diaulus a doctor, now he is an undertaker. What the undertaker now does the doctor too did before.

Epigrams
Book I, epigram XLVII

Zoilus is ill: it is his bed-trappings cause this fever. Suppose him well; what will be the use of scarlet coverlets? What of a mattress from Nile, or of one dipped in strong-smelling purple of Sidon? What but illness displays such foolish wealth? What do you want with doctors? Dismiss all your physicians. Do you wish to become well? Take my bed-trappings.

Epigrams
Book II, epigram XVI

Mather, Cotton
Praemia cum possit medicus!
[The doctor must get his rewards when he can!]

The Angel of Bethesda
Capsila II (p. 11)

McCoy, Dr. Leonard
I'm not a magician, I'm just an old country doctor.

Star Trek
The Deadly Years

I'm a doctor, not an escalator.

Star Trek
Friday's Child

I'm a doctor, not a brick layer.

Star Trek
The Devil In The Dark

I'm a doctor, not a mechanic.

Star Trek
The Doomsday Machine

I'm a doctor, not a coal miner.

Star Trek
The Empath

I'm a doctor, not an engineer.

Star Trek
Mirror, Mirror

What am I, a doctor or a space shuttle conductor?

Star Trek
The Corbomite Maneuver

Melton, John, Sir
. . . the Sunne doth always behold your good successe, and the Earth covers all your ignorances.

Astrologaster (p. 17)

Metcalf, Elizabeth
The most thoughtful doctor I know holds a child's tongue down with a lollipop when he has to look down a small throat.

Reader's Digest
Picturesque Speech and Patter (p. 58)
September 1946

Moore, Thomas
How the Doctor's brow should smile
Crowned with wreaths of camomile; . . .

The Poetical Works of Thomas Moore
Wreaths for Ministers (p. 410)

Nye, Bill
He was a good doctor for horses and blind staggers, but he was out
of his sphere when he strove to fool with the human frame. Change of
scene and rest were favorite prescriptions of his. Most of his patients got
both, especially eternal rest. He made a specialty of eternal rest.

Remarks
My Physician (p. 354)

O'Neill, Eugene
MARY: I hate doctors! they'll do anything—anything to keep you coming
to them. They'll sell their souls! What's worse, they'll sell yours, and you
never know it till one day you find yourself in hell.

Long Day's Journey into Night
Act Two, Scene Two (p. 74)

Osler, Sir William
There are only two sorts of doctors; those who practice with their brains,
and those who practice with their tongues.

Aequanimitas, with Other Addresses
Teaching and Thinking
II (p. 124)

There are two great types of practitioners—the routinist and the
rationalist—neither common in the pure form. Into the clutches of the
demon routine the majority of us ultimately come. The mind, like the
body, falls only too readily into the rut of oft-repeated experiences.
One evening in the far North West, beneath the shadows of the Rocky
Mountains we camped beside a small lake from which diverging in all
directions were deep furrows, each one as straight as an arrow, as far as
the eye could reach. They were the deep ruts or tracks which countless
generations of buffalo had worn in the prairie as they followed each other
to and from the water. In our minds, countless, oft-repeated experiences
wear similar ruts in which we find it easiest to travel, and out of which
many of us never dream of straying.

In Harvey Cushing
The Life of Sir William Osler
Volume I (p. 272)

Owen, John

God and the doctor we alike adore
But only when in danger, not before;
The danger o'er, both are alike requited,
God is forgotten, and the Doctor slighted.

Source unknown
Epigrams

Parrot, Henry

Dacus doth daily to his doctor go,
As doubting if he be in health or no;
For when his friends salute him passing by,
And ask him how he doth in courtesy,
He will not answer thereunto precise,
Till from his doctor he hath ta'en advice.

In William Davenport Adams
English Epigrams
On One who was a Slave to the Physician
cdlxxxiii

Pekkanen, John

Being a good doctor means being incredibly compulsive. It has nothing
to do with flights of intuition or brilliant diagnoses or even saving
lives. It's dealing with a lot of people with chronic diseases that you
really can't change or improve. You can help patients. You can make a
difference in their lives, but you do that mostly by drudgery—day after
day paying attention to details, seeing patient after patient and complaint
after complaint, and being responsive on the phone when you don't feel
like being responsive.

M.D.—Doctors Talk About Themselves
Chapter 3 (p. 70)

Petronius

What good's a doctor but for peace of mind?

The Satyricon
XLII (p. 50)

Pitkin, Walter B.

A country doctor needs more brains to do his work passably than the
fifty greatest industrialists in the world require.

The Twilight of the American Mind
Medicine (p. 118)

Plath, Sylvia

Sunday—the doctor's paradise! Doctors at country clubs, doctors at the seaside, doctors with mistresses, doctors with wives, doctors in church, doctors in yachts, doctors everywhere resolutely being people, not doctors.

The Bell Jar
Chapter Nineteen (p. 276)

Pope, Alexander

Who shall decide, when Doctors disagree,
And soundest Casuists doubt, like you and me?

The Poems of Alexander Pope
Volume III, ii
Epistle iii
To Allen Lord Bathurst, L. 1–2

. . . Banish'd the doctor, and expell'd the friend?

The Poems of Alexander Pope
Volume III, ii
Epistle iii
To Allen Lord Bathurst, L. 330

Prior, Matthew

You tell your doctor, that y' are ill,
And what does he, but write a bill,
Of which you need not read one letter:
The worse the scrawl, the dose the better.
For if you knew but what you take,
Though you recover, he must break.

The Poetical Works of Matthew Prior
Alma
Canto iii, L. 97–102

Proust, Marcel

Inasmuch as a great part of what doctors know is taught them by the sick, they are easily led to believe that this knowledge which patients exhibit is common to them all, and they pride themselves on taking the patient of the moment by surprise with some remarks picked up at a previous bedside.

The Guermantes Way
Part I
My Grandmother's Illness (p. 416)

Proverb
A broken apothecary a new doctor.

In Robert Christy
Proverbs, Maxims and Phrases of All Ages (p. 255)

A doctor is one who kills you to-day to prevent you from dying to-morrow.

In Robert Christy
Proverbs, Maxims and Phrases of All Ages (p. 255)

A doctor's child dies not from disease but from medicine.

In Robert Christy
Proverbs, Maxims and Phrases of All Ages (p. 256)

A loquacious doctor is successful.

In Robert Christy
Proverbs, Maxims and Phrases of All Ages (p. 256)

Proverb, Chinese
Patients worry over the beginning of an illness, doctors worry over its end.

In H.L. Mencken
A New Dictionary of Quotations on Historical Principles

Proverb, French
After death the doctor.

In H.L. Mencken
A New Dictionary of Quotations on Historical Principles

The presence of the doctor is the first part of the cure.

In H.L. Mencken
A New Dictionary of Quotations on Historical Principles

Proverb, German
Good doctors do not like big bottles.

In H.L. Mencken
A New Dictionary of Quotations on Historical Principles

No doctor at all is better than three.

In H.L. Mencken
A New Dictionary of Quotations on Historical Principles

A half doctor near is better than a whole one far away.

In Robert Christy
Proverbs, Maxims and Phrases of All Ages (p. 256)

Proverb, Italian
A doctor and a clown know more than a doctor alone.

> In H.L. Mencken
> *A New Dictionary of Quotations on Historical Principles*

A doctor's mistake is the will of God.

> In H.L. Mencken
> *A New Dictionary of Quotations on Historical Principles*

If the patient dies, the doctor killed him; if he gets well, the saints cured him.

> In H.L. Mencken
> *A New Dictionary of Quotations on Historical Principles*

While the doctor is considering the patient dies.

> In H.L. Mencken
> *A New Dictionary of Quotations on Historical Principles*

Proverb, Latin
Death defies the doctor.

> In H.L. Mencken
> *A New Dictionary of Quotations on Historical Principles*

Before a doctor can cure one patient he must kill ten.

> In H.L. Mencken
> *A New Dictionary of Quotations on Historical Principles*

Proverb, Portuguese
If your friend is a doctor, send him to the house of your enemy.

> In H.L. Mencken
> *A New Dictionary of Quotations on Historical Principles*

Proverb, Russian
Only a fool will make a doctor his heir.

> Source unknown

Proverb, Welsh
Heaven defend me from a busy doctor.

> In H.L. Mencken
> *A New Dictionary of Quotations on Historical Principles*

Rabelais, François
May a hundred devils leap on my body, if there aren't more old drunkards than old doctors!

Gargantua
Book I, Chapter 41 (p. 49)

Richardson, Samuel
. . . when medical men are at a loss what to prescribe, they inquire what their patients best like and forbid them that.

The Works of Samuel Richardson
Volume VII
The History of Clarissa Harlowe
Volume IV
Letter CXVI (p. 460)

Richler, Mordecai
. . . if you ask me, doctors should be taken with a grain of salt. Certain diseases are still a mish-mash to them.

Son of A Smaller Hero
3. Spring 1935 (p. 110)

Romanoff, Alexis Lawrence
Doctors are needed at birth, but are of no use at death.

Encyclopedia of Thoughts
Aphorisms 529

A doctor has to share the suffering of his patients.

Encyclopedia of Thoughts
Aphorisms 2385

The doctor is an angel to an ailing man—
Until the bill appears—beyond his paying plan.

Encyclopedia of Thoughts
Couplets

Each doctor has his own unselfish aim—
On patients' health to have a rightful claim.

Encyclopedia of Thoughts
Couplets

Ruskin, John
They, on the whole, desire to cure the sick, and,—if they are good doctors, and the choice were fairly put to them,—would rather cure their patient and lose their fee, than kill him, and get it.

The Crown of Wild Olives
Work (p. 30)

Shakespeare, William
If thou couldst, doctor, cast
The water of my land, find her disease,
And purge it to a sound and pristine health,
I would applaud thee to the very echo,
That should applaud again.

Macbeth
Act V, Scene III, L. 50

Shaw, George Bernard
Thus, everything is on the side of the doctor. When men die of disease they are said to die from natural causes. When they recover (and they mostly do) the doctor gets the credit of curing them.

The Doctor's Dilemma
Preface on Doctors
The Craze for Operations (p. xv)

Just as the object of a trade union under existing conditions must finally be, not to improve the technical quality of the work done by its members, but to secure a living wage for them, so the object of the medical profession today is to secure an income for the private doctor; and to this consideration all concern for science and public health must give way when the two come into conflict. Fortunately they are always in conflict. Up to a certain point doctors, like carpenters and masons, must earn their living by doing the work that the public wants for them.

The Doctor's Dilemma
Preface on Doctors
Trade Unionism and Science (p. xxxv)

And let no one suppose that the words doctor and patient can disguise from the parties the fact that they are employer and employee.

The Doctor's Dilemma
Preface on Doctors
The Future of Private Practice (p. lxxxi)

Make it compulsory for a doctor using a brass plate to have inscribed on it, in addition to the letters indicating his qualifications, the words, "Remember that I too am mortal."

The Doctor's Dilemma
Preface on Doctors
The Latest Theories (p. xci)

Spence, Sir James Calvert
. . . the doctor needs to know the individual with basic, expert, and specialized understanding if he is to work with success. He sees men of all ages from childhood to senility. He is present at birth and at death. He

observes man in his confidence of full health and in his fear of sickness. He observes him near the noon of day when courage is at its height, and in the small hours of morning when it so often ebbs away. Not only must he understand the individual but he must understand him in many of these variations from the norm.

The Purpose and Practice of Medicine
Chapter 18 (p. 273)

Stanley, Sir Henry Morton
Doctor Livingston, I presume?

On meeting David Livingston in Ujini, Central Africa
November 10, 1871

Sterne, Laurence
. . . there are worse occupations in this world *than feeling a woman's pulse*.

A Sentimental Journey through France and Italy
The Pulse (p. 97)

Stevenson, Robert Louis
Doctors is all swabs.

Treasure Island
Chapter 3 (p. 14)

Stewart, Michael M.
What do you want to be, my son,
When you're all grown up like me?
A doctor like you,
With nothing to do
But handing out pills for a fee.

The New England Journal of Medicine
Manpower Planning (By Degrees) (p. 673)
Volume 287, Number 13, September 28, 1972

Swift, Jonathan
. . . the best Doctors in the World are Doctor *Diet*, Doctor *Quiet*, and Doctor *Merryman*.

The Prose Works of Jonathan Swift
Volume the Fourth
Polite Conversation
Second Conversation (p. 182)

Syrus, Publilius
Male secum agit æger, medicum qui hæredem facit.
[A sick man does ill for himself who makes the doctor his heir.]

Sententiæ
Number 333

Thackeray, William Makepeace
It is not only for the sick man, it is for the sick man's friends that the Doctor comes. His presence is often as good for them as for the patient, and they long for him yet more eagerly.

The History of Pendennis
Chapter LII (p. 597)

Trotter, Wilfred
As long as medicine is an art, its chief and characteristic instrument must be human faculty. We come therefore to the very practical question of what aspects of human faculty it is necessary for the good doctor to cultivate . . . The first to be named must always be the power of attention, of giving one's whole mind to the patient without the interposition of one self. It sounds simple but only the very greatest doctors ever fully attain it. It is an active process and not either mere resigned listening or even politely waiting until you can interrupt. Disease often tell its secrets in a casual parenthesis . . .

Collected Papers (pp. 97–8)

Twain, Mark
He . . . has been a doctor a year now, and has had two patients—no, three, I think; yes, it *was* three. I attended their funerals.

The Guilded Age
Chapter 10 (p. 80)

Unknown
An apple a day keeps the doctor away.

Source unknown

A crafty old doc from Montpelier,
Always said to the girls, "I can heal yer;
  But first please undress
  So I don't have to guess,
But can see yer and smell yer and feel yer."

The Little Limerick Book (p. 6)

There was a young harlot of Clyde
Whose doctor cut open her hide.
  He misplaced his stitches
  And closed the wrong niches;
She now does her work on the side.

In William S. Barring-Gould's
The Lure of the Limerick (p. 137)

A lucky doctor is better than a learned one.

Source unknown

A new doctor, a new grave digger.

Source unknown

A young doctor should have three graveyards.

Source unknown

An ignorant doctor is no better than a murderer.

Source unknown

The king employ'd three doctors daily,
Willis, Heberden, and Baillie;
All exceeding clever men,
Baillie, Willis, Heberden;
But doubtful which most sure to kill is,
Baillie, Heberden, or Willis

In William Davenport Adams
English Epigrams
On George the Fourth's Physicians
cclvxi

From no man yet you've run away!
    Doctor, that may be true;
You've kill'd so many in your day,
    Men mostly fly from you.

In William Davenport Adams
English Epigrams
On a Valiant Doctor
ccxc

[Doctor] A licensed drug pusher.

Esar's Comic Dictionary

[Doctor] A person who never seems willing to let well enough alone.

Esar's Comic Dictionary

Some doctors seem to think that M.D. actually stands for Minor Deity.

Source unknown

Fifty years ago the successful doctor was said to need three things; a top hat to give him Authority, a paunch to give him Dignity, and piles to give him an Anxious Expression.

The Lancet
In England Now (p. 169)
Volume 260, January 20, 1951

Three faces wears the doctor: when first sought
An angel's; and a god's the cure half-wrought;
But when, the cure complete, he seeks his fee,
The Devil looks less terrible than he.

Source unknown

Just what the doctor ordered.

Source unknown

Three doctors are in the duck blind and a bird flies overhead.

The general practitioner looks at it and says, "Looks like a duck, flies like a duck . . . it's probably a duck," and shoots at it but misses and the bird flies away.

The next bird flies overhead, and the pathologist looks at it, then looks through the pages of a bird manual, and says, "Hmmmm . . . green wings, yellow bill, quacking sound . . . might be a duck." He raises his gun to shoot it, but the bird is long gone.

A third bird flies over. The surgeon raises his gun and shoots almost without looking, brings the bird down, and turns to the pathologist and says, "Go see if that was a duck."

Source unknown

If any fatal wretch be sick
Go call the doctor, haste, be quick,
The doctor comes with drop and pill
But don't forget his calomel.

In Louise Pound
American Ballads and Songs
Calomel
Stanza 3 (p. 126)

He sent for doctors for one year
To try their skill.
Dear doctor, your skill's in vain,
There's none like Betsy to save my pain.

In Cecil Sharp
English Folk Songs from the Southern Appalachians
Betsy

Boll Weevil say to de Doctor,
"Better po' out all yo' pills,
When I git through wid de Farmer,
He cain't pay no doctor's bills."

In John Avery Lomax
American Ballads and Folk Songs
The Boll Weevil

Off to the doctor I did go
And I showed him my big toe.

In came the doctor with a bloomin' big lance
Now, young sailor, I'll make you dance!

In came the nurse with a mustard poultice
Banged it on, but I took no notice.

Now I'm well and free from pain
I'll never court flash gals again.

In Stan Hugill
Shanties from the Seven Seas
Banks of the Sacramento

Then comes the doctor, the worst of them all
Saying, "What's been the matter with you all the fall?"
He says that he'll cure you of all your disease.
When your money he's got, you can die if you please.

In Edith Fowke (Editor)
The Penguin Book of Canadian Folk Songs
Hard, Hard Times

You go to the Doctor, you feel mighty ill
The Doctor looks you over, he gives you a pill
Then if you die, they break out the band
The Doctor's done his duty and he doesn't give a damn.

The Book of Navy Songs
Collected by the Trident Society, 1943
Home, Boys, Home

Voltaire

I know nothing more laughable than a doctor who does not die of old age.

Annals of Medical History
Quoted in Pearch Bailey
Voltaire's Relation to Medicine (p. 58)
Volume 1, 1917

Wolfe, Humbert

The doctors are a frightful race.
I can't see how they have the face
to go on practising their base
profession; but in any case
I mean to put them in their place.

Cursory Rhymes
Poems Against Doctors
I

DRAFT

Unknown

Air coming in at a window is as bad as a cross-bow shot.

Source unknown

He that sits with his back to a draft sits with his face to a coffin.

Source unknown

DRUGGIST

Armour, Richard

The druggist is the doctor's friend,
  He serves him with devotion.
For him his shelves and counters bend
  With every pill and potion.
The Druggist works with all his might,
  He gets down early, Monday.
He toils away till late at night;
  His store is open Sunday.
The druggist's quite a pleasant man,
  His jokes are somewhat ribald.
And best, he reads (no other can)
  The words the doctor scribbled.

The Medical Muse
The Druggist

DRUGS

Bierce, Ambrose
Opiate, *n.* An unlocked door in the prison of Identity. It leads into the jail yard.

The Devil's Dictionary

Dunne, Finley Peter
"Don't ye iver take dhrugs?" asked Mr. Hennessy.

"Niver whin I'm well," said Mr. Dooley. "Whin I'm sick, I'm so sick I'd take annything."

Mr. Dooley Says
Drugs (p. 99)

Glasow, Arnold
Miracle drug: One you can afford.

Quote, The Weekly Digest
August 27, 1967

Jackson, James
It is my own practice to avoid drugs as much as possible; and I more frequently find it difficult to persuade people from using them, than to induce them to take them. But I hope that you will not believe me to be distrustful of the power of drugs to do real service to the sick, under proper circumstances. I am far otherwise. And in reference to this point, I wish to tell you that your success in the use of medicines may depend somewhat on the temper with which you give them. You must be hopeful and feel an interest in them. Do not, like a cold stepfather, leave them to make their own way in the world; but watch them in their course.

Letters to a Young Physician

Unknown
A drug is a substance which when injected into a guinea pig produces a scientific paper.

Source unknown

EAR WAX

Flaubert, Gustave

Cerumen. Human wax. Should not be removed: it keeps insects from entering the ears.

Dictionary of Accepted Ideas

ELECTROCARDIOGRAM

Kraus, Jack
ELECTROCARDIOGRAM: ticker tape.

Quote, The Weekly Digest
February 5, 1967 (p. 117)

The medical profession is the only one in which anybody
professing to be a physician is at once trusted,
although nowhere else is an untruth more dangerous.
Pliny the Elder – (See p. 272)

ENEMA

Armour, Richard
I hymn the lowly enema,
   A very humble type, O,
Not half so bright or half so keen
   Or sharp as is the hypo.

. . .

No thing of glitter, poised in air,
   No instrument of grace,
There's this about the enema
   At least: it has its place.

The Medical Muse
Friends and Enemas

EPIDEMIC

Bierce, Ambrose
EPIDEMIC, *n*. A disease having a sociable turn and few prejudices.

The Enlarged Devil's Dictionary

[Cold] An ailment for which there are many unsuccessful remedies, with whiskey being the most popular.
Unknown – (See p. 48)

EPIDERMIS

Bierce, Ambrose

EPIDERMIS, *n*. The thin integument which lies immediately outside the skin and immediately inside the dirt.

The Enlarged Devil's Dictionary

ERROR

Amiel, Henri Frédéric
An error is the more dangerous in proportion to the degree of truth which it contains.

Journal Intime
December 26, 1852 (p. 43)

Beard, George M.
. . . as quantitative truth is of all forms of truth the most absolute and satisfying, so quantitative error is of all forms of error the most complete and illusory.

Popular Science Monthly
Experiments with Living Human Beings (p. 751)
Volume 14, 1879

Jefferson, Thomas
Error of opinion may be tolerated where reason is left free to combat it.
Inaugural Address of the President of the United States

Latham, Peter Mere
Amid many possibilities of error, it would be strange indeed to be always in the right.

In William B. Bean
Aphorisms from Latham (p. 19)

Osler, Sir William
. . . errors in judgment must occur in the practice of an Art which consists largely in balancing probabilities . . .

Teacher and Student (p. 19)

ESOPHAGUS

Bierce, Ambrose
ESOPHAGUS, *n*. That portion of the alimentary canal that lies between
pleasure and business.

The Enlarged Devil's Dictionary

EXAMINATION

Holmes, Oliver Wendell

You will remember, of course, always to get the weather-gage of your patient. I mean, to place him so that the light falls on his face and not on yours. It is a kind of ocular duel that is about to take place between you; you are going to look through his features into his pulmonary and hepatic and other internal machinery, and he is going to look into yours quite as sharply to see what you think about his probabilities for time or eternity.

Medical Essays
The Young Practitioner (p. 387)

Mayo, Charles H.

Examination must be within reason for the sick, or near-sick, and its extent will be based on the judgment and experience of the physician.

Minnesota Medicine
When Does Disease Begin? Can This Be Determined by Health Examination?
Volume 15, January 1932

Mayo, William J.

Sometimes I wonder whether today we take sufficient care to make a thorough physical examination before our patient starts off on the round of the laboratories, which have become so necessary that oftentimes we do not fully appreciate the value of our five senses in estimating the condition of the patient.

Collected Papers of the Mayo Clinic & Mayo Foundation
Discussion of Paper by T.E. Keys
Volume 30, 1938

The examining physician often hesitates to make the necessary examination because it involves soiling the finger.

Journal-Lancet
The Cancer Problem
Volume 35, July 1, 1915

EXPERIENCE

Adams, Henry Brooks
All experience is an arch to build upon.

The Education of Henry Adams
Rome (p. 87)

Bernard, Claude
Speaking concretely, when we say "making experiments or making observations," we mean that we devote ourselves to investigation and to research, that we make attempts and trials in order to gain facts from which the mind, through reasoning, may draw knowledge or instruction.

Speaking in the abstract, when we say, "relying on observation and gaining experience," we mean that observation is the mind's support in reasoning, and experience the mind's support in deciding, or still better, the fruit of exact reasoning applied to the interpretation of facts.

Observation, then, is what shows facts; experiment is what teaches about facts and gives experience in relation to anything.

An Introduction to the Study of Experimental Medicine
Part I, Chapter I, Section II (p. 11)

Bowen, Elizabeth
Experience isn't interesting till it begins to repeat itself—in fact, till it does that, it hardly *is* experience.

Death of the Heart
The World (p. 14)

Cardozo, Benjamin N.
Often a liberal antidote of experience supplies a sovereign cure for a paralyzing abstraction built upon a theory.

The Paradoxes of Legal Science
Chapter IV (p. 125)

da Vinci, Leonardo

Experience is never at fault; it is only your judgment that is in error in promising itself such results from experience as are not caused by our experiments. For having given a beginning, what follows from it must necessarily be a natural development of such a beginning, unless it has been subject to a contrary influence, while, if it is affected by any contrary influence, the result which ought to follow from the aforesaid beginning will be found to partake of this contrary influence in a greater or less degree in proportion to the said influence is more or less powerful than the aforesaid beginning.

Leonardo da Vinci's Notebooks (pp. 56–7)

Hoffmann, Friedrich

In medicine there are two supports—experience, which is the first parent of truth; and reason, which is the key to medical science. Experience comes first in order, and reason follows. Hence in medical affairs reasons which are not founded on experience have no value.

Fundamenta Medicinae
Physiology
Chapter I, 7 (p. 5)

Latham, Peter Mere

Wherefore then serveth *experience*, and of what use is it? Its first and best use is for the guidance of him that has it. Its next, and hardly less important use, is that it enables him to judge rightly the experience of others.

In William B. Bean
Aphorisms from Latham (p. 93)

Nothing is so difficult to deal with as man's own Experience, how to value it according to amount, what to conclude from it, and how to use it and do good with it.

In William B. Bean
Aphorisms from Latham (p. 94)

Unknown

Where did you get your good judgment? Answer: From my experience.

Where did you get your experience? Answer: From my *poor* judgment.

Source unknown

von Baeyer, Adolf

Men who are capable of modifying their first beliefs are very rare. This ability was one of the reasons for the success of Claude Bernard and Pasteur. Out of a very vivid imagination they forged new hypotheses all the time but abandoned them with equal ease as soon as experience contradicted them.

In Richard Willstätter
From My Life
Chapter 6 (p. 141)

Wilde, Oscar

DUMBY: Experience is the name everyone gives to their mistakes.

The Works of Oscar Wilde
Lady Windermere's Fan
Third Act (p. 60)

EXPERIMENT

Bernard, Claude
Considered in itself, the experimental method is nothing but reasoning by whose help we methodically submit our ideas to experience—the experience of fact.

An Introduction to the Study of Experimental Medicine
Introduction (p. 2)

Latham, Peter Mere
Experiment is like a man traveling to some far off place, and finding no place by the way where he can sit down and rest himself, and few or no guide posts to tell him whether he be in the right direction for it or not. Still he holds on. Perhaps he has been there before, and is pretty sure of this being the direction in which he found it. Or, perhaps he has never been there, but some of his friends have, and they told him of this being the right road to it. And so it may be that, by his own sagacity and the help of well-informed friends, he reaches it at last. Or, after all his own pains, and all his friends can do for him, it may be that he never reaches it at all.

In William B. Bean
Aphorisms from Latham (p. 91)

Planck, Max
Experimenters are the shocktroops of science.

Scientific Autobiography (p. 110)

An experiment is a question which science poses to Nature, and a measurement is the recording of Nature's answer.

Scientific Autobiography (p. 110)

Weyl, Hermann

Allow me to express now, once and for all, my deep respect for the work of the experimenter and for his fight to wring significant facts from an inflexible Nature, who says so distinctly "No" and so indistinctly "Yes" to our theories.

The Theory of Groups and Quantum Mechanics (p. xx)

A 4-year review of trauma admissions . . . reveals that 2.5% of such admissions were due to being struck by falling coconuts . . .
Peter Barss – (See p. 179)

FACT

Bernard, Claude
If the facts used as a basis for reasoning are ill-established or erroneous, everything will crumble or be falsified; and it is thus that errors in scientific theories most often originate in errors of fact.

An Introduction to the Study of Experimental Medicine
Part I, Chapter I, Section III (p. 13)

Facts are neither great nor small in themselves.

An Introduction to the Study of Experimental Medicine
Part I, Chapter II, Section III (p. 34)

A fact is nothing in itself, it has value only through the idea connected with it or through the proof it supplies.

An Introduction to the Study of Experimental Medicine
Part I, Chapter II, Section VII (p. 53)

Crothers, Samuel McChord
The trouble with facts is that there are so many of them.

The Gentle Reader
That History Should be Readable (p. 183)

Fabing, Howard
Marr, Ray
Facts are not science—as the dictionary is not literature.

Fischerisms (p. 21)

Hare, Hobart Amory
At first it is impossible for the novice to cast aside the minor symptoms, which the patient emphasises as his major ones, and to perceive clearly that one or two facts that have been belittled in the narration of the story of the illness are in reality the stalk about which everything in the case must be made to cluster.

Practical Diagnosis
Introduction (p. 17)

Hooker, Worthington
The physician who narrows his view down to a certain set of facts, is in danger of becoming enamored of them. And if he does, he is straightway in the fog and mists of error. He forsakes the practical for a fruitless will o' the wisp pursuit of the ideal, all the while believing that he has found vast mines of truth, and very confident that his search is to be still more abundantly rewarded.

Lessons from the History of Medical Delusions (p. 35)

Husserl, E.
Merely fact-minded sciences make merely fact-minded people.

The Crisis of European Sciences and Transcendental Phenomenology
Part I, Section 1 (p. 6)

Latham, Peter Mere
Bear in mind then, that abstractions are *not facts*; and next bear in mind that *opinions* are not facts.

In William B. Bean
Aphorisms from Latham (p. 36)

Mayo, William J.
The man of science in searching for the truth must ever be guided by the cold logic of facts, and be animated by scientific imagination.

Collected Papers of the Mayo Clinic & Mayo Foundation
Perception
Volume 20, 1928

Nightingale, Florence
What you want are facts, not opinions—

Notes on Nursing
Observing the Sick (p. 105)

FEE

Clowes, William
Before you meddle with him, make your bargain wisely, now he is in pain, for he is but a bad pay-master, and therefore follow this rule: *Get your money while he's ill, for when he's well you never will*.

Selected Writings of William Clowes
A Tragical History (p. 63)

de Mondeville, Henri
Never dine with a patient who is in your debt, but get your dinner at an inn, otherwise he will deduct his hospitality from your fee.

In Samuel Evans Massengill
A Sketch of Medicine and Pharmacy (p. 307)

Dunne, Finley Peter
I wondher why ye can always read a doctor's bill an' ye niver can read his purscription?

Mr. Dooley Says
Drugs (pp. 93–4)

Fielding, Henry
So little did their doctors delight in death that they discharged the corpse after a single fee.

Tom Jones
Book II, Chapter 9 (p. 34)

Graves, Richard
Three doctors, met in consultation,
Proceed with great deliberation;
The case was desperate, all agreed,
But what of that? they must be fee'd.
They write then (as't was fit thay shou'd)

But for their own, not patients' good.
Consulting wisely (don't mistake, sir)
Not what to give, but what to take, sir.

In William Davenport Adams
English Epigrams
The Consultation
cclxxxii

Hazlitt, William Carew
A Physitian demanded money of another for one of his patients that was dead long before. He was answered that it was a worke of charity to visit the sick; but if he was so earnest for money, the only way was for him to visit the dead, and then he would never want money more.

Shakespeare Jest Books
Volume III
Conceit, Clichés, Flashes and Whimzies
Number 176

Hood, Thomas
The doctor look'd and saw the case
Plain as the nose *not* on his face.
"O! hum—ha—yes—I understand."
But then arose a long demur,
For not a finger would he stir
Till he was paid his fee in hand;
That matter settled, there they were,
With Hunks well strapp'd upon his chair.

The Poetical Works of Thomas Hood
Volume 1
A True Story
Stanza 12

McElwee, Tom
Cardiologist's Fee—Heart-earned money.

Quote, The Weekly Digest
June 2, 1968 (p. 437)

Unknown
You say, without reward or fee
   Your uncle cured me of a dangerous ill;
I say, he never did prescribe for me;
   The proof is plain—I'm living still.

In William Davenport Adams
English Epigrams
On a Doctor
dccclxx

"Better to roam the fields, for health unbought,
Than fee the doctor for a nauseous draught."
This maxim long I happily pursued,
And fell disease my health then ne'er pursued;
But to be more than well at length I tried—
The doctor came at last—and then I died!

In William Davenport Adams
English Epigrams
On Medical Advice
Epistle "To My Honoured Kinsman John Driden"
cmii

A physician's fees are ill-gotten gains.

In Evan Esar
20,000 Quips and Quotes

THAT'S CERTAINLY A BIG CAVITY YOU HAVE HERE~
HAVE HERE... HAVE   H E R E ...

[Cavity] A hollow place in a tooth
ready to be stuffed with a dentist's bill.
Unknown – (See p. 39)

FEVER

Colman, George (the Younger)
The doctor look'd wise:—"A slow fever," he said:
Prescribed sudorifics,—and going to bed.
"Sudorifics in bed," exclaim'd Will, "are humbugs!
I've enough of them there, without paying for drugs!"

In Helen and Lewis Melville's
An Anthology of Humorous Verse
Lodgings for Single Gentlemen

Herold, Don
The sweetest words of tongue or pen: The child is 98.6 again.

The Happy Hypochondriac (p. 12)

Kipling, Rudyard
An'd that blasted English drizzle wakes the fever in my bones; . . .

Mandalay

Cure them if they have fever, but by no means work charms.

Kim
XII (p. 303)

MacFadden, Bernarr
If you feed a cold, as often done, you frequently have to starve a fever.

Physical Culture
When a Cold is Needed
February 1934

Milton, John
Fever, the eternal reproach to the physicians.

The Reason of Church-Government
Preface

Osler, Sir William

Humanity has but three great enemies: fever, famine and war; of these by far the greatest, by far the most terrible, is fever.

In Harvey Cushing
The Life of Sir William Osler
Volume I (p. 435)

Paulos, John Allen

Consider a precise number that is well known to generations of parents and doctors: the normal human body temperature of 98.6 Fahrenheit. Recent investigations involving millions of measurements reveal that this number is wrong; normal human body temperature is actually 98.2 Fahrenheit. The fault, however, lies not with Dr. Wunderlich's original measurements—they were averaged and sensibly rounded to the nearest degree: 37 Celsius. When this temperature was converted to Fahrenheit, however, the rounding was forgotten and 98.6 was taken to be accurate to the nearest tenth of a degree. Had the original interval between 36.5 and 37.5 Celsius been translated, the equivalent Fahrenheit temperatures would have ranged from 97.7 to 99.5. Apparently, discalculia can even cause fevers.

A Mathematician Reads the Newspaper
Ranking Health Risks: Experts and Laymen Differ (p. 139)

Proverb, Persian

Show him death, and he'll be content with fever.

Source unknown

Ransom, John Crowe

Here lies a lady of beauty and high degree.
Of chills and fever she dies, of fever and chills,
The delight of her husband, her aunts, an infant of three,
And of medicos marveling sweetly on her ills.

The Poetry of John Crowe Ransom
Here Lies a Lady (p. 64)

Shakespeare, William

He had a fever when he was in Spain,
And when the fit was on him, I did mark
How he did shake; . . .

Julius Caesar
Act I, Scene II, L. 119–121

FLU

Beacock, Cal
A bunch of germs were whooping it up
In the Bronchial Saloon.
The bacillus handling the larynx
Was jazzing a gig-time tune,
While back of the tongue in a solo game
Sat Dangerous Ah Kerchoo.
And watching his luck was his light of his love
The malady known as Flu.

Reader's Digest
The Pundit
January 1986

GALL BLADDER

Rogers, Will
Then he turned and exclaimed with a practiced and well-subdued enthusiasm, "It's the Gall-Bladder—just what I was afraid of." Now you all know what the word "afraid of," when spoken by a doctor, leads to. It leads to more calls.

The Autobiography of Will Rogers
Chapter Twelve (p. 153)

GENERAL PRACTITIONER

Hubbard, Kin

Th' thing I like about general practitioners is that you don't have t' let 'em know a week ahead when you're goin' t' be sick.

Abe Martin: Hoss Sense and Nonsense (p. 94)

GERM PLASM

Hirsch, N.D.M.

The germ cells are like immortal princes confined in isolated castles that they themselves have built; in other castles princesses are confined and they no more than the princes are satisfied with their sister cells. So both princes and princesses ignite their castles by the flame of love; which although ultimately destructive of the existing castle, builds mightier and higher ones from their very flames and ashes.

The germ plasm has lived millions of years at least; it is perpetually experiencing, for all life, as distinguished from inorganic matter, endures—lives with and through the past. In our own lives we are the total of all our past. In our own lives we are the total of all our past experiences . . . we are ever growing older like the universe of the immortal Bergson. So too the germ-plasm; it is literally the heir of all the ages of *germ-plasm* experience, and the germ plasm is living, is experiencing as much as we and in like manner. The germ cells have sexual yearnings, pugnacious tendencies, and *they create; mutations are their darlings* . . .

Genius and Creative Experience (pp. 60 and 61)

GOD

Bonner, W.B.

It seems to me highly improper to introduce God to solve our scientific problems.

In Charles-Albert Reichen
A History of Astronomy (p. 100)

Feynman, Richard

God was invented to explain mystery. God is always invented to explain those things that you do not understand. Now, when you finally discover how something works, you get some laws which you're taking away from God; you don't need him anymore. But you need him for the other mysteries. So therefore you leave him to create the universe because we haven't figured that out yet; you need him for understanding those things which you don't believe the laws will explain, such as consciousness, or why you only live to a certain length of time—life and death—stuff like that. God is always associated with those things that you do not understand. Therefore I don't think that the laws can be considered to be like God because they have been figured out.

In P.C.W. Davies and J. Brown
Superstrings: A Theory of Everything (p. 208)

Paré, Ambroise

Je le pensay, et Dieu le guarit
[I dressed his wound, and God healed it.]

In Oliver Wendell Holmes
Medical Essays
The Medical Profession in Massachusetts (p. 365)

GOUT

Ray, John

With respect to gout, the physician is a lout.

A Complete Collection of English Proverbs (p. 35)

Unknown

The French have taste in all they do,
  Which we are quite without;
For Nature, which to them gave *goût*,
  To us gave only gout.

In Harvey Cushing
The Life of Sir William Osler
Volume I (p. 576)

GYNECOLOGY

Rivers, Joan
The gynecologist says, "Relax, relax, I can't get my hand out, relax." I wonder why I'm not relaxed. My feet are in the stirrups, my knees are in my face, and the door is open facing me . . . And my gynecologist does jokes "Dr. Schwartz at your cervix!" "I'm dilated to meet you!" "Say ahhh." "There's Jimmy Hoffa!" There's no way you can get back at that son of a bitch unless you learn to throw your voice.

In Roz Warren
Glibquips (p. 70)

A male gynecologist is like an auto mechanic who has never owned a car.

Snow, Carrie
Source unknown

Unknown
The eminent authorities who study the geography
Of this obscure but interesting land
Are able to indulge a taste for feminine topography
And view the graphic details close at hand.

Source unknown
The Doctor's Lament

HEADACHE

Ray, John

When the *head* aches all the body is the worse.

A Complete Collection of English Proverbs (p. 12)

Twain, Mark

Do not undervalue the headache. While it is at its sharpest it seems a bad investment; but when relief begins, the unexpired remainder is worth four dollars a minute.

Following the Equator
Pudd'nhead Wilson's New Calendar (p. 517)

HEALING

Bonaparte, Napoleon
You know, my dear doctor, the art of healing is simply the art of lulling and calming the imagination.

In J. Christopher Herold (Editor)
The Mind of Napoleon
Science and the Arts (p. 140)

Butler, Samuel
Dogs with their tongues their wounds do heal,
But men with hands, as thou shalt feel.

Hudibras
Part I, Canto ii, L. 773–774

Eddy, Mary Baker
Then comes the question, how do drugs, hygiene and animal magnetism heal? It may be affirmed that they do not heal, but only relieve suffering temporarily, exchanging one disease for another.

Science and Health (p. 483)

Matthew 4:23
And Jesus went about all Galilee, teaching in their synagogues, and preaching the gospel of the kingdom, and healing all manner of sickness and all manner of disease among the people.

The Bible

Mencken, H.L.
. . . it will never get well if you pick it.

The American Mercury
What is Going on in the World (p. 257)
Volume XXX, Number 119, November 1933

Psalms 147:3
He healeth the broken in heart, and bindeth up their wounds.

The Bible

Shakespeare, William

What wound did ever heal but by degrees?

Othello
Act II, Scene III

Unknown

One is not so soon healed as hurt.

Source unknown

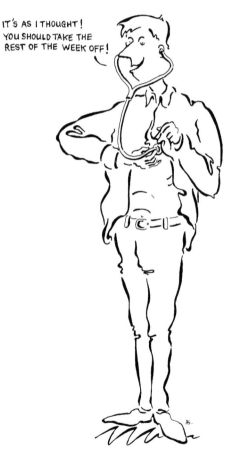

Physician, heal thyself.
Luke 4:23 – (See p. 267)

HEALTH

Cervantes, Miguel de
The beginning of health lies in knowing the disease and in the sick man's willingness to take the medicines which the physician prescribes . . .

Don Quixote
Part II, Chapter 60 (p. 392)

Chesterton, Gilbert Keith
The trouble about always trying to preserve the health of the body is that it is so difficult to do without destroying the health of the mind.

Come to Think of It
On the Classics

da Vinci, Leonardo
Strive to preserve your health; and in this you will the better succeed in proportion as you keep clear of the physicians, for their drugs are a kind of alchemy concerning which there are no fewer books than there are medicines.

Leonardo da Vinci's Notebooks (p. 65)

Donleavy, J.P.
. . . when you don't have any money, the problem is food. When you have money, it's sex. When you have both, it's health . . .

The Ginger Man
Chapter 5 (p. 39)

Eddy, Mary Baker
Health is not a condition of matter, but of Mind; nor can the material senses bear reliable testimony on the subject of health.

Science and Health (p. 120)

159

Eliot, T.S.

The sense of wellbeing! It's often with us
When we are young, but then it's not noticed;
And by the time one has grown to consciousness
It comes less often.

The Elder Statesman
Act II (p. 54)

Emerson, Ralph Waldo

The first wealth is health. Sickness is poor-spirited, and cannot serve any one; it must husband its resources to live. But health or fullness answers its own ends, and has to spare, runs over, and inundates the neighborhoods and creeks of other men's necessities.

Complete Works
Volume 6
Conduct of Life
Power

Fuller, Thomas

Sickness is felt, but health not at all.

Gnomologia
4160

Herophilus

To lose one's health renders science null, art inglorious, strength effortless, wealth useless and eloquence powerless.

In Samuel Evans Massengill
A Sketch of Medicine and Pharmacy (p. 28)

Holmes, Oliver Wendell

If you mean to keep as well as possible, the less you think about your health the better.

Over the Teacups
Chapter viii (p. 186)

La Rochefoucauld, François de

Preserving the health by too severe a rule is a wearisome malady.

Maxims of La Rochefoucauld (p. 157)

Proverb

Health and sickness surely are men's double enemies.

In George Herbert
Outlandish Proverbs
#1011

Ray, John
Early to go to bed, and early to rise, makes a man healthy, wealthy, and wise.

A Complete Collection of English Proverbs (p. 33)

Health is better than wealth.

A Complete Collection of English Proverbs (p. 120)

Health without money is half a sickness.

A Complete Collection of English Proverbs (p. 12)

Romains, Jules
Healthy people are sick people who don't know it.

Knock
Act 1 (p. 12)

Romanoff, Alexis Lawrence
One who does not care about his own health, or life, will soon be either disabled or dead.

Encyclopedia of Thoughts
Aphorisms 1033

Sacks, Oliver
Health is infinite and expansive in mode, and reaches out to be filled with the fullness of the world; whereas disease is finite and reductive in mode, and endeavors to reduce the world to itself.

Awakenings
Perspectives (p. 209)

Shaw, George Bernard
Use your health, even to the point of wearing it out. That is what it is for. Spend all you have before you die; and do not outlive yourself.

The Doctor's Dilemma
Preface on Doctors
The Latest Theories (p. xcii)

Simmons, Charles
A man too busy to take care of his health is like a mechanic too busy to take care of his tools.

Laconic Manual and Brief Remarker (p. 234)

He that wants health wants everything.

Laconic Manual and Brief Remarker (p. 234)

Thomson, James
Health is the vital Principle of Bliss,
And Exercise of Health.

The Castle of Indolence
Canto II, Stanza lvii

Twain, Mark
He had had much experience of physicians, and said "the only way to keep your health is to eat what you don't want, drink what you don't like, and do what you'd druther not."

Following the Equator
Pudd'nhead Wilson's New Calendar (p. 459)

Unknown
By the side of sickness health becomes sweet.

Source unknown

Good health is above wealth.

Source unknown

Health and understanding are the two great blessings of life.

Source unknown

Health is happiness.

Source unknown

He that goes to bed thirsty rises healthy.

Source unknown

He that would be healthy must wear his winter clothes in summer.

Source unknown

He who has not health has nothing.

Source unknown

Health dwells with the peasant.

Source unknown

Health is not valued till sickness comes.

Source unknown

We are usually the best men when in the worst health.

Source unknown

Without health life is not life, life is lifeless.

Source unknown

[Health] The greatest of all blessings, but only when you're sick.

Esar's Comic Dictionary

Walton, Izaak
. . . look to your health: and if you have it, praise God, and value it next to a good conscience; for health is the second blessing that we mortals are capable of; a blessing that money cannot buy.

The Compleat Angler
Part I, Chapter 21 (p. 224)

HEART

Barnard, Christiaan N.

. . . it is infinitely better to transplant a heart "than to bury it so it can be devoured by worms."

Time
People (p. 36)
October 31, 1969

Barnes, Djuna

We are adhering to life now with our last muscle—the heart.

Nightwood
La Somnambule (p. 50)

De Bakey, Michael E.

If you can think of how much love there would be in hundreds of hearts, then that is how much love there is in a plastic heart. When you grow up you will understand how very much love that is.

Newsweek
Heart of the Matter (p. 56)
June 6, 1966

Harvey, William

The heart, consequently, is the beginning of life; the sun of the microcosm. Even as the sun in his turn might well be designated the heart of the world; for it is the heart by whose virtue and pulse the blood is moved, perfected, made apt to nourish, and is preserved from corruption and coagulation; it is the household divinity which, discharging its function, nourishes, cherishes, quickens the whole body, and is indeed the foundation of life, the source of all action. But of these things we shall speak more opportunely when we come to speculate upon the final cause of this motion of the heart.

An Anatomical Disquisition on the Motion of the Heart and Blood in Animals
Chapter 8 (p. 286)

Hellerstein, Herman
Coronary heart disease is a silent disease and the first manifestation frequently is sudden death.

Newsweek
Tests to Avoid Attack (p. 64)
August 6, 1984

Legrain, G.
The heart is a god . . . the stomach is its chapel.

Répertoire Généalogique Et Onomastique Du Musée Du Caire
Statues et Statuettes
III
42225, e, I, II (p. 60)

. . . he [Physician] is the only person with whom one dares to talk continually of oneself . . .
Hannah More – (See p. 268)

HIPPOCRATIC OATH

Hippocrates

I swear by Apollo the physician, and Æsculapius, and Health, and All-heal, all the gods and goddesses, that, according to my ability and judgment, I will keep this Oath and this stipulation—to reckon him who taught me this Art equally dear to me as my parents, to share my substance with him, and relieve his necessities if required; to look upon his offspring in the same footing as my own brothers, and to teach them this art, if they shall wish to learn it, without fee or stipulation; and that by precept, lecture, and every other mode of instruction, I will impart knowledge of the Art to my own sons, and those of my teachers, and to disciples bound by a stipulation and oath according to the law of medicine, but to none others. I will follow that system of regimen which, according to my ability and judgment, I consider for the benefit of my patients, and abstain from whatever is deleterious and mischievous. I will give no deadly medicine to anyone if asked, nor suggest any such counsel; and in like manner I will not give to a woman a pessary to produce abortion. With purity and with holiness I will pass my life and practice my Art. I will not cut persons laboring under the stone, but will leave this to be done by men who are practitioners of this work. Into whatever houses I enter, I will go into them for the benefit of the sick, and will abstain from every voluntary act of mischief and corruption; and, further from the seduction of females or males, of freemen and slaves. Whatever, in connection with my professional practice, or not in connection with it, I see or hear, in the life of men, which ought not to be spoken abroad, I will not divulge, as reckoning that all such should be kept secret. While I continue to keep this Oath unviolated, may it be granted to me to enjoy life and the practice of the art, respected by all men, in all times! But should I trespass and violate this Oath, may the reverse be my lot.

The Oath

HIVES

Eisenschiml, Otto

The Indian medicine man used weird chants and dances to mystify his tribe. The medicine man of today uses cryptic cabalas in his prescription, and long Latin words for simple diseases. This impresses the patient, who in turn relishes to regale his visitor with expressions they cannot understand and are ashamed to have explained to them. After all, it does make a difference whether you suffer from angioneurotic edema or only have the hives.

The Art of Worldly Wisdom
Part Eleven (p. 135)

HOSPITAL

Beckett, Samuel
What sky! What light! Ah in spite of all it is a blessed thing to be alive in such weather, and out of hospital.

All that Fall (p. 9)

Bernard, Claude
. . . I consider hospitals only as the entrance to scientific medicine; they are the first field of observation which a physician enters; but the true sanctuary of medical science is a laboratory; only there can he seek explanations of life in the normal and pathological states by means of experimental analysis.

Experimental Medicine
Part II, Chapter II, Section X (p. 140)

Bevan, Aneurin
I would rather be kept alive in the efficient if cold altruism of a large hospital than expire in a gush of warm sympathy in a small one.

Speech, April 30, 1946
House of Commons

Bierce, Ambrose
HOSPITAL, *n.* A place where the sick generally obtain two kinds of treatment—medical by the doctor and inhuman by the superintendent.

The Enlarged Devil's Dictionary

Browne, Sir Thomas
. . . for the world, I count it not an Inne, but an Hospitall; and a place, not to live, but to die in.

Religio Medici
Part II, Section 11 (p. 95)

Browning, Elizabeth Barrett
I think it frets the saints in heaven to see
How many desolate creatures on the earth
Have learnt the simple dues of fellowship
And social comfort, in a hospital.

Aurora Leigh
Third Book (p. 110)

Compton-Burnett, Ivy
I suppose I shall subscribe to hospitals. That's how people seem to give
to the poor. I suppose the poor are always sick. They would be if you
think.

A Family and A Fortune
Chapter 4 (p. 122)

Kerr, Jean
One of the most difficult things to contend with in a hospital is the
assumption on the part of the staff that because you have lost your gall
bladder you have also lost your mind.

Please Don't Eat the Daises
Operation operation (p. 143)

Kraus, Jack
Hospital: Ail House

Quote, The Weekly Digest
April 2, 1967 (p. 277)

Mayo, William J.
The hospital should be a refuge to which the sick might go for relief as
they went before our Savior, . . .

Journal of the Michigan Medical Society
The Teaching Hospital of the University of Michigan
Volume 25, January 1926

Ray, John
A suit of *law* and an urinal brings a man to the hospital.

A Complete Collection of English Proverbs (p. 14)

Southerne, Thomas
. . . when, wee'r worn,
Hack'd, hewn with constant service, thrown aside
To rust in peace; or rot in Hospitals.

The Loyal Brother
Act 1, Scene I (p. 8)

Thomson, James
. . . lo! A goodly Hospital ascends;
In which they bade each human Aid be nigh,
That could the Sick-Bed smoothe of that unhappy Fry.

It was a worthy edifying Sight,
And gives to Human-Kind *peculiar* Grace,
To see kind Hands attending Day and Night,
With tender Ministry, from Place to Place.
Some prop the Head; some, from the pallid Face,
Wipe off the faint cold Dews weak Nature sheds;
Some reach the healing Draught: the whilst, to chase
The Fear supreme, around their soften'd Beds,
Some holy Man by Prayer all opening Heaven dispreds.

The Castle of Indolence
Canto II, Stanzas lxxiv–lxxv

Unknown
[Hospital] Bedpan alley

Esar's Comic Dictionary

[Hospital] Where a patient has a few friends and many enemas.

Esar's Comic Dictionary

[Hospital] A place where the food usually tastes worse than the medicine.
Esar's Comic Dictionary

HOSPITAL: An institution in which the lucky patients are healed and released, while the unlucky ones are kept and subjected to unnecessary drugs, unnecessary surgery and unmitigated expense.

In Richard Iannelli
The Devil's New Dictionary

HYPOCHONDRIAC

Ace, Goodman
If you're a hypochondriac, first class, you awaken each morning with the firm resolve not to worry; everything is going to turn out all wrong.

The Fine Art of Hypochondria
Who Am I (p. 13)

Askey, Vincent
When it comes to your health, I recommend frequent doses of that rare commodity among Americans—common sense. We are rapidly becoming a land of Hypochondriacs, from the ulcer-and-martini executives in the big city to the patent medicine patrons in the sulphur-and-molasses belt.

The Land of Hypochondriacs
Address, October 20, 1960
Bakersfield, California

Colton, Charles Caleb
Those hypochondriacs, who, like Herodius, give up their whole time and thoughts to the care of their health, sacrifice unto life, every noble purpose of living; striving to support a frail and feverish being here, they neglect an hereafter; they continue to patch up and repair their mouldering tenement of clay, regardless of the immortal tenant that must survive it; agitated by greater fears than the apostle, and supported by none of his hopes, they 'die daily.'

Lacon
1.139

Hypochondriacs squander large sums of time in search of nostrums by which they vainly hope they may get more time to squander.

Lacon
2.70

Cvikota, Clarence
Hypochondriac: Pill collector

Quote, The Weekly Digest
April 7, 1968 (p. 277)

Friedman, Shelby
My hypochondriac wife takes so many timed disintegrating capsules, you can hear her ticking.

Quote, The Weekly Digest
March 10, 1968 (p. 197)

Herold, Don
Even a hypochondriac can have appendicitis.

The Happy Hypochondriac (p. 16)

Karch, Carroll S.
Hypochondriac—Enjoying pill health.

Quote, The Weekly Digest
September 15, 1968 (p. 217)

Ogutsch, Edith
Hypochondriac: A person of ill repute.

Quote, The Weekly Digest
May 7, 1967 (p. 377)

Thomson, James
And moping here did *Hypochondria* sit,
Mother of Spleen, in Robes of various Dye,
Who vexed was full oft with ugly fit,
And some her frantic deem'd, and some her deem'd a wit.

A lady proud she was of ancient blood,
Yet oft her fear her pride made crouchen low,
She felt or fancy'd in her fluttering mood,
All the diseases which at the spittles know,
And sought all physic which the shops bestow,
And still new leaches and new drugs would try,
Her humour ever wavering to and fro,
For sometimes she would laugh and sometimes cry,
Then sudden waxed wroth, and all she knew not why.

The Castle of Indolence
Hypochondria
Stanzas 75–76

Unknown
[Hypochondriac] The man who has everything.

Esar's Comic Dictionary

[Hypochondriac] A person whose affliction is a fiction.

Esar's Comic Dictionary

[Hypochondriac] A person who never lets his health stand in the way of his ailments.

Esar's Comic Dictionary

[Hypochondriac] A person who spends half his time meditating on his body and the other half medicating it.

Esar's Comic Dictionary

A hypochondriac never gets cured of any disease until he acquires another.

In Evan Esar
20,000 Quips and Quotes

HYPODERMIC NEEDLE

Battles, William Snowden
For many hold 'twould be so hard
  Through Heaven's gate to wheedle
A doctor as to drive a camel through
  A hypodermic needle.

In Mary Lou McDonough
Poet Physician
The Doctor's Dream (p. 81)

Chekov and Sulu
"The needle won't hurt, Chekov. Take off your shirt, Chekov. Roll over, Chekov. Breathe deeply, Chekov. Blood sample, Chekov. Marrow sample, Chekov. Skin sample, Chekov. If . . . if I live long enough . . . I'm going to run out of samples."

"Oh, you'll live."

"Ah, yes, but I won't enjoy it!"

Star Trek
The Deadly Years

Kernan, F.C.
Hypodermic needle: Sick shooter.

Quote, The Weekly Digest
March 19, 1967 (p. 237)

ILL

Adams, Cedric

From the day, so to speak, of creation
  Mankind has been subject to ills,
Though he waited for civilization
  To learn about powders and pills.
The pains that so often assailed him
  Made our primitive ancestor squirm,
And he never found out that what ailed him
  Was only a germ.
An ache, under certain conditions,
  Has been through the ages the same,
But the efforts of learned physicians
  Have fitted it out with a name,
We are told with mellifluous quickness
  Which no one can quite understand
That every conceivable sickness
  Is due to a gland.
Our forefathers suffered acutely
  From things that afflict us today,
But we are aware more minutely
  Of what is the trouble than they.
The doctors are endlessly clever
  At telling us habits to shun—
Being sick is as painful as ever,
  But vastly more fun.

Poor Cedric's Almanac (pp. 129–30)

Butler, Samuel

. . . I reckon being ill as one of the great pleasures of life, provided one is not too ill and is not obliged to work till one is better.

The Way of All Flesh
LXXX (p. 363)

175

Gide, André
Those who have never been ill are incapable of real sympathy for a great many misfortunes.

Journals
Volume 3
July 25, 1930 (p. 279)

Trilling, Lionel
We are all ill: but even a universal sickness implies an idea of health.

The Liberal Imagination
Art and Neurosis (p. 174)

von Ebner-Eschenbach, Marie
Imaginary ills belong to the incurable.

Aphorisms (p. 24)

It is not the mortal but the incurable illnesses which are the worst.

Aphorisms (p. 44)

Wilder, Thornton
For what human ill does not dawn seem to be an alleviation?

The Bridge of San Luis Rey
Part 3 (p. 119)

ILLNESS

Bond, J.
Bond, S.
A simple distinction we need to make here is between *illness* and *disease*. Disease refers to a medical concept of pathology, which is indicated by a group of signs and symptoms. The presence or absence of a disease, as indicated by signs and symptoms, is clinically defined by the medical profession. The doctor or his substitute, using a common body of knowledge, makes the decision as to whether or not a person has a disease. In contrast, illness is defined by the person who had the signs and symptoms. It refers primarily to a person's subjective experience of 'health' and 'ill-health' and is indicated by the person's reactions to the symptoms.

Sociology and Health Care
Chapter 8 (p. 200)

Chekhov, Anton
If doctors prescribe too many remedies for an illness it probably means that the illness can't be cured at all.

The Cherry Orchard
Act 1 (p. 21)

Rutherford, D.
Amiga-Ga : Brain failure, characterized by regression of speech to a high-speed beeping noise.

Atari-Kiri : Extreme reaction to repeatedly poor video scores.

Nitendo-itis: Painful bruising of the thumbs from frantic button pressing.

Segalepsy : Epileptic fits triggered by flickering video images.

See-3DO : Retinal burnout from viewing monitor screens too closely.

Newsweek
Letter
October 18, 1993

Södergran, Edith
I lie all day and wait for night,
I lie all night and wait for day.

We Women
Days of Sickness

Sontag, Susan
Illness is the night-side of life, a more onerous citizenship. Everyone who is born holds dual citizenship, in the kingdom of the well and in the kingdom of the sick. Although we all prefer to use only the good passport, sooner or later each of us is obligated, at least for a spell, to identify ourselves as citizens of that other place.

Illness as Metaphor
Chapter 1 (p. 3)

Fatal illness has always been viewed as a test of moral character, but in the nineteenth century there is a great reluctance to let anybody flunk the test.

Illness as Metaphor
Chapter 5 (p. 41)

Unknown
It takes longer for a person to get over an illness if compensation sets in.

In Evan Esar
20,000 Quips and Quotes

Welty, Eudora
He did not like illness, he distrusted it, as he distrusted the road without signposts.

Selected Stories of Eudora Welty
Death of a Traveling Salesman (p. 232)

Woolf, Virginia
There is, let us confess it (and illness is the great confessional), a childish outspokenness in illness; things are said, truths blurted out, which the cautious respectability of health conceals.

The Moment
On Being Ill (p. 13)

Considering how common illness is, how tremendous the spiritual change that it brings, how astonishing, when the lights of health go down, the undiscovered countries that are then disclosed, what wastes and deserts of the soul a slight attack of influenza brings to view, what precipices and lawns sprinkled with bright flowers a little rise of temperature reveals, what ancient and obdurate oaks are uprooted in us by the act of sickness . . .

The Moment
On Being Ill (p. 9)

INJURY

Barss, Peter
A 4-year review of trauma admissions to the Provincial Hospital, Alotau, Milne Bay Province, reveals that 2.5% of such admissions were due to being struck by falling coconuts. Since mature coconut palms may have a height of 24 up to 35 meters and an unhusked coconut may weigh 1 to 4 kg, blows to the head of a force exceeding 1 metric ton are possible.

The Journal of Trauma
Injuries Due to Falling Coconuts
Abstract (p. 990)
Volume 24, Number 11, 1984

INSOMNIAC

Crichton-Browne, Sir James

A lady of my acquaintance, after a slight operation, suffered from insomnia which drugs failed to relieve. Her doctor, however, was very resourceful and on visiting her on the Sunday forenoon said: "I see there's a sermon on the wireless at eight this evening . . ." There is no soporific better than a dry sermon.

From the Doctor's Notebook
Insomnia (p. 202)

Unknown

A person who sleeps better with a pill than with a pillow.

In Evan Esar
Esar's Comic Dictionary

INTERNIST

Unknown
A doctor who specializes in general practice.

In Evan Esar
Esar's Comic Dictionary

You can never use too much KY.

Source unknown

INTESTINE

Dunne, Finley Peter

. . . though I have patches on me pantaloons, I've ne'er a wan on me intestines.

Mr. Dooley's Opinions
Thanksgiving (p. 127)

INVALID

Mansfield, Katherine

Nearly all my improved health is pretence—acting. What does it amount to? . . . I am an absolutely helpless invalid. What is my life? It is the existence of a parasite.

The Journal of Katherine Mansfield
October 10, 1922 (p. 252)

Proverb, Irish

Every invalid is a physician.

Source unknown

Proverb, Italian

Why live like an invalid to die as a healthy man?

Source unknown

Yourcenar, Marguerite

Nothing seemed simpler: a man has the right to decide how long he may usefully live . . . sickness disgusts us with death, and we wish to get well, which is a way of wishing to live. But weakness and suffering, with manifold bodily woes, soon discourage the invalid from trying to regain ground: he tires of those respites which are but snares, of that faltering strength, those ardors cut short, and that perpetual lying in wait for the next attack.

Memoirs of Hadrian
Patientia (pp. 280, 281)

IT

Carroll, Lewis

". . . the patriotic archbishop of Canterbury, found it advisable—"

"Found *what*?" said the Duck.

"Found *it*," the Mouse replied, rather crossly: "of course you know what 'it' means."

"I know what 'it' means well enough, when I find a thing," said the Duck: "it's generally a frog, or a worm."

The Complete Works of Lewis Carroll
Alice's Adventures in Wonderland

Weingarten, Violet

. . . report for a routine checkup feeling like a hypochondriac bacause obviously you're in perfect health, and the doctor mumbles something about "it" having to come out, no rush, next week will be plenty of time.

Intimations of Mortality (p. 3)

JOURNALS

Stuart, Copans A.
Why, dear colleagues, must our findings
Now be put in sterile bindings?
Once physicians wrote for recreation.
Our great teachers through the ages,
Fracastro, and other sages,
Found writing could be fun, like fornication . . .

Perspectives in Biology and Medicine
Winter 1973 (p. 232)

KIDNEY

Smith, Homer
There are those who say that the human kidney was created to keep the blood pure, or more precisely, to keep our internal environment in an ideal balanced state. This I must deny. I grant that the human kidney is a marvelous organ, but I cannot grant that it was purposefully designed to excrete urine or to regulate the composition of the blood or to subserve the physiological welfare of *Homo sapiens* in any sense. Rather I contend that the human kidney manufactures the kind of urine that it does, and it maintains the blood in the composition which that fluid has, because the kidney has a certain functional architecture; and it owes that architecture not to design or foresight or to any plan, but to the fact that the earth is an unstable sphere with a fragile crust, to the geologic revolutions that for six hundred million years have raised and lowered continents and seas, to the predacious enemies, and heat and cold, and storms and droughts; to the unending succession of vicissitudes that have driven the mutant verterbrates from sea into fresh water, into desiccated swamps, out upon the dry land, from one habitation to another, perpetually in search of the free and independent life, perpetually failing, for one reason or another, to find.

From Fish to Philosopher
Chapter 13 (pp. 210–11)

Unknown
There was a young fellow named Sydney,
Who drank till he ruined his kidney.
  It shriveled and shrank,
  As he sat there and drank,
But he'd had a good time at it, hadn't he?

In Louis Untermeyer
Lots of Limericks (p. 143)

KIDNEY: A bean-shaped organ used primarily as a model for swimming pools. Its secondary use is in the human body, where it collects urine, a fact which many pool owners, particularly those with a lot of children, find unpleasantly appropriate.

In Richard Iannelli
The Devil's New Dictionary

Piss clear, and defy the physician.
John Ray – (See p. 275)

KING'S EVIL

Shakespeare, William
MACD: What's the disease he means?
MAL: 'Tis call'd the evil;
A most miraculous work in this good king;
Which often, since my here-remain in England,
I have seen him do. How he solicits Heaven,
Himself best knows; but strangely-visited people,
All swoln and ulcerous, pitiful to the eye,
The mere despair of surgery, he cures,
Hanging a golden stamp about their necks,
Put on with holy prayers. And 'tis spoken,
To the succeeding royalty he leaves
The healing benediction.

Macbeth
Act IV, Scene III, L. 147–158

LIFE

Byron, Lord George Gordon
'Tis very certain the desire of life
Prolongs it: this is obvious to the physicians,
When patients, neither plagued with friends nor wife,
Survive through very desperate conditions . . .

Don Juan
Canto II, Stanza LXIV

LIVER

Jonson, Ben
OVERDO: . . . the lungs of the tobacconist are rotted, the liver spotted, the brain smoked like the backside of the pig-woman's booth, here, and the whole body within, black as her pan you saw e'en now, without.

Bartholomew Fair
Act II, Scene VI (p. 84)

Molière
GERONTE: It seems to me you are locating them wrongly: the heart is on the left and the liver is on the right.

SGANARELLE: Yes, in the old days that was so, but we have changed all that, and we now practice medicine by a completely new method.

The Reluctant Doctor
Act 2, Scene 4

Shakespeare, William
If he were opened, and you find so much blood in his liver as will clog the foot of a flea, I'll eat the rest of the anatomy.

Twelfth Night
Act III, Scene II, L. 68–69

Unknown
A strip-teaser in Fall River
Caused a sensitive fellow to quiver.
  The esthetic vibration
  Brought soulful elation.
Besides, it was good for his liver.

In Louis Untermeyer
Lots of Limericks (p. 169)

LUNGS

Ott, Susan
Roses are red
Violets are blue
Without your lungs
Your blood would be too.

The New England Journal of Medicine
A Pulmonologist's Valentine (p. 739)
Volume 304, Number 12, 1981

MALADY

Hawthorne, Nathaniel
Some maladies are rich and precious and only to be acquired by the right of inheritance or purchased with gold.

Mosses from an Old Manse: The Procession of Life (p. 199)

Maturin, Charles R.
A malady
Preys on my heart that med'cine cannot reach.

Bertram
Act IV, Scene II (p. 52)

MAL DE MER

Byron, Lord George Gordon
The best of remedies is a beef-steak
Against sea-sickness; try it, sir, before
You sneer, and I assure you this is true,
For I have found it answer—so may you.

Don Juan
Canto II, Stanza XIII

Farris, Jean
Mal de mer—An ocean-motion notion.

Quote, The Weekly Digest
August 4, 1968 (p. 97)

Flaubert, Gustave
Sea-sickness. To avoid, all you have to do is think of something else.

Dictionary of Accepted Ideas

Sterne, Laurence
. . . the cells are broke loose one into another, and the blood, and the lymph, and the nervous juices, with the fix'd and volatile salts, are all jumbled into one mass—good g—! everything turns round in it like a thousand whirlpools . . .

Tristram Shandy
Volume VII, Chapter II (p. 4)

MAN

Byron, Lord George Gordon
Man is a carnivorous production,
And must have meals, at least one meal a day;
He cannot live, like woodcocks, upon suction,
But, like the shark and tiger, must have his prey;
Although his anatomical construction
Bears vegetables, in a grumbling way,
Your laboring people think beyond all question,
Beef, veal, and mutton better for digestion.

Don Juan
Canto II, Stanza LXVII

Koestler, Arthur
But, glory be, man is not a flat-earth dweller all the time—only most of the time. Like the universe in which he lives, he is in a state of continuous creation.

The Act of Creation
Chapter XX (p. 363)

Marquis, Don
the supercilious silliness
of this poor wingless bird
is cosmically comical
and stellarly absurd.

the lives and times of archy & mehitabel
archys life of mehitabel
archy turns revolutionist (p. 227)

Pascal, Blaise
. . . what is man in nature? A nothing in comparison with the infinite, an all in comparison with the nothing, a mean between nothing and everything. Since he is infinitely removed from comprehending the extremes, the ends of things and the beginnings are hopelessly hidden

from him in an impenetrable secret; he is equally incapable of seeing the nothing of which he is made, and the infinite in which he is swallowed up.

Pensées
Section II, 72

Pickering, James Sayre
Man is a freak, colloidal combination of thirteen elements which happen to have a chemical affinity for each other, and is the strangest and one of the most amusing accidents of nature. The market value of the substance of the average man is about 98c.

The Stars are Yours (p. 230)

MEASLES

Proverb, Chinese
Starve the measles and nourish the small-pox.

Source unknown

Ward, Artemus
Did you ever have the measles, and if so, how many?

The Complete Works of Artemus Ward
The Census (p. 69)

MEDICAL

Ace, Goodman
A rule of thumb in the matter of medical advice is to take everything any doctor says with a grain of aspirin.

The Fine Art of Hypochondria
Only Sick People Go To the Doctors (p. 44)

Armstrong, John
For want of timely care
Millions have died of medicable wounds.

Art of Preserving Health
Book III, L. 515 (p. 92)

da Costa, J. Chalmers
A medical man in the plumage of pretense resembles the humming bird, which, when stripped of its plumage, is not larger than the bumble bee.

The Trials and Triumphs of the Surgeon
Stepping Stones and Stumbling Blocks
Part III (p. 230)

Dickens, Charles
There might be medical doctors a-cocking their medical eyes.

Tale of Two Cities
Book 3, Chapter IX

Esar, Evan
It's surprising how many medical authorities you can find over bridge tables.

20,000 Quips and Quotes

The most difficult task of the medical profession is to train patients to become sick during office hours only.

20,000 Quips and Quotes

Medicine is a science, acquiring a medical practice an art.

20,000 Quips and Quotes

Hewitt, Barnard
Still I wish I knew a half a dozen good long medical terms to give an authentic air of learning to my conversation.

The Doctor in Spite of Himself
Act III (p. 67)

Hoffmann, Friedrich
In physics experience can best be sought from mathematics and mechanics, chemistry, and anatomy; in medical practice experience derives most abundantly from the observations of diseases, and from more accurate histories and cures.

Fundamenta Medicianae
Physiology
Chapter I, 8 (p. 5)

To live in a medical fashion, that is, according to the strict and academic rules of the physicians, is to live miserably and uncomfortably.

Fundamenta Medicianae
Medical Hygiene
Chapter I, 10 (p. 103)

Morris, Joseph F.
Medical reporter—Staph writer.

Quote, The Weekly Digest
July 21, 1968 (p. 57)

Neigher, Harry
They were gabbing, at the Pen & Pencil, about a medical specialist in Greenwich who is so suspicious of late he gets the feeling someone's listening in on his stethoscope.

Connecticut Sunday Herald
February 19, 1967

Renard, Jules
There is nothing so sickening as to leaf through a medical dictionary.

In Evan Esar
20,000 Quips and Quotes

Shaw, George Bernard
It is not the fault of our doctors that the medical service of the community, as at present provided for, is a murderous absurdity. That any sane nation, having observed that you could provide for the supply of bread by giving bakers a pecuniary interest in baking for you, should go on and give a surgeon a pecuniary interest in cutting off your leg, is enough to make one despair of political humanity.

The Doctor's Dilemma
Preface on Doctors (p. v)

Steinbeck, John
The medical profession is unconsciously irritated by lay knowledge.

East of Eden
Chapter 54, Section 1 (p. 589)

SHE MUST BE EXPECTING
TO GET FLU!

For fever, take one teaspoon of salt mixed in water.
After that put a spoonful of salt inside each stocking
as soon as you feel a chill coming on.
Unknown – (See p. 304)

MEDICAL MNEMONICS

Unknown
Robert Taylor Drinks Cold Beer
R=roots, T=trunks, D=divisions, C=cords, B=branches

The Brachial Plexus
Source unknown

A mnemonic which will allow you to remember whether a nerve is sensory (S), motor (M) or both (B). Each word corresponds to the cranial nerve of the same number:
Some Surgeons Make Money But My Brother Says Buxom Blondes Make More.

Source unknown

Never Lower Tillie's Pants. Grandmother Might Come Home.
Navicular
Lunate
Triquetral
Pisiform
Greater Multangular
Lesser Multangular
Capitate
Hamate

Mnemonic for the bones of the wrist
Source unknown

MEDICAL TERMINOLOGY

Unknown

Artery: The study of fine paintings.

Bacteria: Back door to cafeteria.

Barium: What you do when C.P.R. fails.

Benign: What you are after you be eight.

Bowel: A letter like A, E, I, O, U.

Caesarean Section: A district in Rome.

Cat Scan: Searching for kitty.

Cauterize: Made eye contact with her.

Colic: A sheep dog.

Coma: A punctuation mark.

Congenita: Friendly.

D & C: Where Washington is.

Dilate: To live longer.

Enema: Not a friend.

Fester: Quicker.

Genital: Non Jewish.

G. I. Series: Baseball games between soldiers.

Grippe: A suitcase.

Hangnail: A coat hook.

High Colonic: Jewish religious holiday.

Medical Staff: A doctor's cane.

Minor Operation: Coal digging.

Morbid: A higher offer.

Nitrate: Lower than the day rate.

Node: Was aware of.

Organic: Church musician.

Outpatient: A person who has fainted.

Pap Smear: Fatherhood test.

Pelvis: Cousin of Elvis.

Post-Operative: A letter carrier.

Prostate: Flat on your back.

Protein: In favor of young people.

Recovery Room: Place to do upholstery.

Rectum: Dang near killed him.

Rheumatic: Amorous.

Secretion: Hiding anything.

Seizure: Roman emperor.

Tablet: A small table.

Terminal: Getting sick at the airport.

Tibia: Country in North Africa.

Tumor: An extra pair.

Urine: Opposite of you're out.

Varicose: Near by.

Vein: Conceited.

Back-Woods Medical Terminology
Source unknown

MEDICINE

Bacon, Francis

. . . the poets did well to conjoin music and medicine in Apollo, because the office of medicine is but to tune this curious harp of man's body and to reduce it to harmony.

Advancement of Learning
Second Book, X, 2

Medicine is a science which hath been, as we have said, more professed than laboured, and yet more laboured than advanced: the labour having been in my judgment, rather in circle than in progression.

Advancement of Learning
Second Book, X, 3

Surely every medicine is an invention; and he that will not apply new remedies must expect new evils.

Bacon's Essays
Of Innovations (p. 65)

Berkenhout, John

I do not deny that many lives might be saved by the skillful administration of proper medicine; but a thousand indisputable facts convince me, that the present established practice of physic in England is infinitely destructive of the lives of his Majesty's subjects. I prefer the practice of old women, because they do not sport with edged tools; being unacquainted with the powerful articles of the *Materia Medica*.

In Roy Porter
The Greatest Benefit to Mankind (p. 262)

Bloom, Samuel W.

Art and science march hand in hand through the history of medicine, each taking turns at the lead. A century or more ago science became the dominant partner and has remained so ever since. The art of medicine meanwhile, like an honored but neglected wife, walked behind, passive and obedient to call at those odd moments when the master needed a change of pace.

The Doctor and His Patient
Chapter I (p. 33)

Bonaparte, Napoleon

Medicine is not an exact and positive science but a science based on conjectures and observations. I would have more confidence in a physician who has not studied the natural sciences than in one who has.

In J. Christopher Herold (Editor)
The Mind of Napoleon
Science and the Arts (p. 139)

Brown, John

It is in medicine as in the piloting of a ship—rules may be laid down, principles expounded, charts exhibited; but when a man has made himself master of all these, he will often find his ship among breakers and quicksands, and must at last have recourse to his own craft and courage.

Horae Subsecivae

Bryce, John

Medicine, the only profession that labours incessantly to destroy the reason for its own existence.

Address
At dinner for General W.C. Gorgas, March 23, 1914

Butler, Samuel

Learn'd he was in medic'nal lore,
For by his side a pouch he wore,
Replete with strange hermetic powder
That wounds nine miles point-blank would solder.

Hudibras
Part I, Canto II, L. 223–226

"'Tis not amiss, e're giv'n o'er,
To try one desp'rate Med'cine more"
For where your Case can be no worse,
The desp'rat'st is the wisest course.

Hudibras
An Historical Epistle of Hudibras to Sidrophel
L. 5–8

Cervantes, Miguel de
. . . God who sends the wound sends the salve.

Don Quixote
Part II, Chapter 19 (p. 262)

Charles, Prince of Wales
The whole imposing edifice of modern medicine is like the celebrated tower of Pisa—slightly off balance.

Quoted in
The Observer
January 2, 1983

Chekhov, Anton
Medicine is my lawful wife and literature is my mistress. When I get tired of one I spend the night with the other.

Letters on the Short Story
Letter to A.S. Souvorin, September 11, 1888 (p. 42)

Churchill, Winston
I have been inclined to feel from time to time that there ought to be a hagiology of medical science and that we ought to have saints' days to commemorate the great discoveries which have been made for all mankind, and perhaps for all time—or for whatever time may be left to us. Nature, like many of our modern statesmen, is prodigal of pain. I should like to find a day when we can take a holiday, a day of jubilation, when we can fête good Saint Anaesthesia and chaste and pure Saint Antiseptic.

Speech
Guildhall, London, September 10, 1947
In F.B. Czarnomski
The Wisdom of Winston Churchill (p. 59)

I do not profess to be very deeply acquainted with the science of medicine. I am not a surgeon myself. My experiences in medicine have been vivid and violent, and completely absorbing while they were going on. Nevertheless, I cannot claim that they have given me that broad, detached, general experience which, I believe, is the foundation for all correct scientific action.

Speech
March 2, 1944
Royal College of Physicians

Colman, George (the Younger)
When taken,
To be well shaken.

Broad Grins
The Newcastle Apothecary, Stanza 12

Croll, Oswald
The choice also of the Medicines must always be considered, and their preparations and compositions made by the Physitian himself, and not carelessly left to others. He is truly a genuine Physitian who can tell how (not only by Reason, as mear Rationall Physitians doe, but) by their own hand to prepare the medicaments . . .
Philosophy Reformed and Improved in Four Profound Tractates (pp. 151–2)

Crookshank, F.G.
Medicine is to-day an Art or Calling, to whose exercise certain Sciences are no doubt ancillary; but she has forfeited pretension to be deemed a Science, *because* her Professors and Doctors decline to define fundamentals or to state first principles, and to refuse to consider, in express terms, the relations between Things, Thoughts and Words involved in their communications to others.
In C.K. Ogden and I.A. Richards
The Meaning of Meaning
Supplement II
The Importance of a Theory of Signs and a Critique of Language in the Study of Medicine (p. 33)

de Guevara, Antonio
Medicine is to be praised when it is in the hands of a Phisition that is learned, grave, wise, stayed and of experience . . .
The Familiar Epistles of Sir Anthonie of Guevara
Of Seven Notable Benefits Proceeding from the Good Physition (p. 285)

de Madariaga, Salvadore
There is no medicine; there are only medicine-men.
Essays with a Purpose
On Medicine (p. 172)

Dickinson, Emily
It knew no Medicine—
It was not Sickness—then—
Nor any need of Surgery—
And therefore—'twas not Pain—

The Complete Poems of Emily Dickinson
#559

Is Heaven a physician?
They say that He can heal;
But medicine posthumous
Is unavailable.

The Complete Poems of Emily Dickinson
#1270

Drake, Daniel
Medicine is not a science of meditations, but of observation.

An Introductory Lecture, on the Means of Promoting the
Intellectual Improvement of the Students (p. 13)

Dryden, John
Better to hunt in Fields, for Health unbought,
Than fee the Doctor for a nauseous Draught.
The Wise, for Cure, on Exercise depend;
God never made his Work for Man to mend.

The Poems of John Dryden
Volume IV
To John Dryden, of Chesterton
L. 92–95 (p. 1532)

Ecclesiasticus 38:4
The most high hath created medicines out of the earth, and a wise man
will not abhor them.

The Bible

Esar, Evan
The practice of medicine has advanced so much in recent years that it is
now impossible for a doctor not to find something wrong with you.

20,000 Quips and Quotes

Franklin, Benjamin
Many dishes, many diseases. Many medicines, few cures.

Poor Richard
1734

. . . he is the best doctor who knows the worthlessness of the most
medicines.

In Logan Clendening
Modern Methods of Treatment
Part I, Chapter II (p. 20)

Garth, Samuel
The Patient's Ears remorseless he assails,
Murthers with Jargon where his Med'cine fails.

The Dispensary
Canto II, L. 96

Goethe, Johann Wolfgang von
The spirit of Medicine can be grasped with ease;
Study the great and little world, my friend,
To let it all go in the end
As God may please!

Faust
The First Part of the Tragedy
The Study (2) (p. 95)

Haggard, Howard W.
Mystery, magic and medicine: in the beginning they were one and the
same.

Mystery, Magic and Medicine (p. 9)

Herophilus
Medicines are nothing in themselves, if not properly used, but the very
hands of the gods, if employed with reason and prudence.

In Samuel Evans Massengill
A Sketch of Medicine and Pharmacy (p. 28)

Heschel, Abraham J.
Medicine is more than a profession. Medicine has a soul, and its calling
involves not only the application of knowledge and the exercise of skill
but also facing a human situation. It is not an occupation for those to
whom career is more precious than humanity or for those who value
comfort and serenity above service to others. The doctor's mission is
prophetic.

The Insecurity of Freedom
The Patient as a Person (p. 28)

Hoffmann, Friedrich
As far as medicine uses the principles of physics, it can be properly called
a science; as far as it relies on practice, it can be called an art.

Fundamenta Medicianae
Physiology
Chapter I, 9 (p. 6)

Frequent changes of medicines proclaims the ignorance of the physician
and is calamitous for the patients.

Fundamenta Medicianae
Therapeutics
Chapter I, 37 (p. 137)

Holmes, Oliver Wendell
I firmly believe that if the whole *materia medica* could be sunk to the bottom of the sea, it would be all the better for mankind and all the worse for the fishes.

Address to the Massachusetts Medical Society
May 30, 1860

Hulme, Keri
"What is your objection to hospitalisation and treatment?" The doctor is curious but dispassionate.

"Primarily, that I forgo control over myself and my destiny. Secondly, medicine is in a queer state of ignorance. It knows a lot, enough to be aware that it is ignorant, but practitioners are loath to admit that ignorance to patients. And there is no holistic treatment. Doctor does not confer with religious who does not confer with dietician who does not confer with psychologist . . . "

"What you are saying basically is that you have no trust in doctors or current medicine?"

"Right on."

The Bone People
IV, 12, (pp. 415–6)

Huxley, Thomas Henry
Medicine was the foster-mother of Chemistry, because it has to do with the preparation of drugs and the detection of poisons; of Botany, because it enabled the physician to recognise medicinal herbs; of Comparative Anatomy and Physiology, because the man who studied Human Anatomy and Physiology for purely medical purposes was led to extend his studies to the rest of the animal world.

Collected Essays
Volume III
Science and Education (p. 213)

Jerome, Jerome K.
. . . I never read a patent medicine advertisement without being impelled to the conclusion that I am suffering from the particular disease therein dealt with in its most virulent form.

Three Men in a Boat
Chapter 1 (p. 2)

Kipling, Rudyard
To discuss medicine before the ignorant is of one piece with teaching the peacock to sing . . .

Kim
XII (p. 313)

Kraus, Karl
Medicine—"Your money and your life!"

Half-Truths & One-and-a-Half Truths (p. 111)

Lyly, John
Oh ye Gods, have ye ordeyned for every malady a medicine, for every sore a salve, for every paine a plaster . . .

Euphues (p. 61)

Martin, Walter
The very success of medicine in a material way may now threaten the soul of medicine. Medicine is something more than the cold mechanical application of science to human disease. Medicine is a healing art. It must deal with individuals, their fears, their hopes and their sorrows. It must reach back further than a disease that the patient may have to those physical and emotional environmental factors which condition the individual for the reception of disease.

Inaugural address as president of the American Medical Association
News report of June 23, 1954

Mather, Cotton
. . . the angelical conjunction of medicine with divinity.

In Harvey Cushing
The Life of Sir William Osler
Volume I (p. 69)

Mayo, Charles H.
Medicine can be used only as people are educated to its accomplishments.

Collected Papers of the Mayo Clinic & Mayo Foundation
International Medical Progress
Volume 23, 1931

It would seem that the study of medicine does not always contribute to broadmindedness, as men who choose medicine as a profession are apt to lose rather than gain breadth of perception. It could be said rather that medicine develops individualism.

Medical Life
The Value of Broadmindedness
Volume 34, April 1927

Mencken, H.L.
The aim of medicine is surely not to make men virtuous: it is to safeguard and rescue them from the consequences of their vices. The true physician does not preach repentance; he offers absolution.

In Herbert V. Prochnow and Herbert V. Prochnow, Jr
A Treasury of Humorous Quotations

Montaigne, Michel de
The arts that promise to keep our bodies and souls in health promise a great deal; withal, there are none that less keep their promise.

Essays
Book the Third
Chapter 13 (p. 524)

Osler, Sir William
. . . medicine, unlike law and theology, is a progressive science . . .

In Harvey Cushing
The Life of Sir William Osler
Volume I (p. 129)

The desire to take medicine is perhaps the greatest feature which distinguishes man from animal.

In Harvey Cushing
The Life of Sir William Osler
Volume I (p. 342)

The practice of medicine is an art, not a trade; a calling, not a business; a calling in which your heart will be exercised equally with your head. Often the best part of your work will have nothing to do with potions and powders, but with the exercise of an influence of the strong upon the weak, of the righteous upon the wicked, of the wise upon the foolish. To you, as the trusted family counselor, the father will come with his anxieties, the mother with her hidden grief, the daughter with her trials, and the son with his follies. Fully one-third of the work you do will be entered in other books than yours. Courage and cheerfulness will not only carry you over the rough places of life, but will enable you to bring comfort and help to the weak-hearted and will console you in the sad hours when, like Uncle Toby, you have "to whistle that you may not weep."

The Master-Word in Medicine
Part III (pp. 29–30)

The critical sense and sceptical attitude of the Hippocratic school laid the foundation of modern medicine on broad lines, and we owe to it: *first*, the emancipation of medicine from the shackles of priestcraft and of caste; *secondly*, the conception of medicine as an art based on accurate observation, and as a science, an integral part of the science of man and of nature; *thirdly*, the high moral ideals, expressed in that "most memorable of human documents" (Gomperz), the Hippocratic oath; and *fourthly*, the conception and realization of medicine as the profession of a cultivated gentleman.

Aequanimitas
Chauvinism in Medicine (p. 266)

In no profession does culture count for so much as in medicine.

Student Life (p. 113)

Ovid
For the sharp medic'cine is the patient's cure.

The Art of Love
Amores
Book III, Elegy XI, L. 21

. . . the same medicine will both harm and cure me.

Tristia
Book II, L. 20

Medicine sometimes removes, sometimes bestows safety; showing what plant is healthful, what harmful.

Tristia
Book II
L. 269

The healing art knows not how to remove crippling gout, it helps not the fearful dropsy.

Ex Ponto
Book I, iii

Proust, Marcel
For, medicine being a compendium of the successive and contradictory mistakes of medical practitioners, when we summon the wisest of them to our aid, the chances are that we may be relying on a scientific truth the error of which will be recognized in a few years' time.

The Guermantes Way
Part I
My Grandmother's Illness (p. 409)

Proverb, Spanish
El tiempo cura el enfermo, no el unguento.
[Time, and not medicine, cures the sick.]

Source unknown

Psalms 147:1
He healeth those that are broken in heart: and giveth medicine to heal their sickness.

The Bible

Romains, Jules
Medicine is a rich soil but it doesn't yield its harvest unaided.

Knock
Act 1 (p. 11)

Romanoff, Alexis Lawrence
The desire to live is the best medicine of all.

Encyclopedia of Thoughts
Aphorisms 2048

Good medicine is man's salvation;
Excessive use gives aggravation.

Encyclopedia of Thoughts
Couplets

Seneca
What physician can heal his patient on a flying visit?.

Ad Lucilium Epistulae Morales
Volume I
Epistle xl, Section 4

It is medicine, not scenery, for which a sick man must go a-searching.

Ad Lucilium Epistulae Morales
Volume III
Epistle civ, Section 18

. . . not even medicine can master incurable diseases.

Ad Lucilium Epistulae Morales
Volume III
Epistle xciv, Section 24

No good physitian who dispares to cure.

Clemency
Book I (p. 30)

Shakespeare, William
By mediciane life may be prolonged, yet death
Will seize the doctor too.

Cymbeline
Act V, Scene V, L. 29–30

Great griefs, I see medicine the less.

Cymbeline
Act IV, Scene II, L. 220

Out, loathed medicine! hated potion, hence!

A Midsummer-Night's Dream
Act III, Scene II, L. 264

Not poppy, nor mandragora,
Nor all the drowsy syrups of the world,
Shall ever medicine thee to that sweet sleep
Which thou ow'dst yesterday.

Othello
Act III, Scene III, L. 331–334

Smollett, Tobias

. . . Sir the practice of medicine is one of the most honourable professions exercised among the sons of men; a profession which hath been revered at all periods and in all nations, and even held sacred in the most polished ages of antiquity.

Sir Launcelot Greaves
Chapter XXIV (p. 192)

Szasz, Thomas

Formerly, when religion was strong and science weak, men mistook magic for medicine; now, when science is strong and religion weak, men mistake medicine for magic.

The Second Sin
Science and Scientism (p. 115)

Unknown

It is all right to talk about practicing what you preach, but doctors prescribe lots of medicine that they don't have to take themselves.

Reflections of a Bachelor

Medicines are not meant to live on.

Source unknown

Patients are simples that grow in every medical man's garden.

Source unknown

PRESCRIPTION: Simple but effective tincture of kindness. Compound tincture of thoughtfulness and consideration. The stimulating elixir of pleasant manner and good humor. The soothing ointment of sympathy and understanding. For the severe pain of doubt and fear which plagues the patient in the unfamiliar atmosphere of a hospital for the first time, there is no better analgesic than a few easily swallowed capsules of patience and reassurance.

CAUTION: This medication has been marketed for some time under the abbreviated trade name of TLC. This is, Tender Loving Care. If allowed to stand on the shelf too long unused, it becomes bitter and unpleasant to take. This had led patients to misinterpret the initials to mean Total Lack of Concern.

DOSAGE: Virtually unlimited. To date, no annoying sensitivities have been reported.

American Journal of Nursing
Old-Fashioned Remedy (p. 696)
March 1968

Young, Edward
Will toys amuse, when med'cines cannot cure?

Night Thoughts
Night II, L. 67 (p. 21)

MENSTRUATION

Butler, Brett
I would like it if men had to partake in the same hormonal cycles to which we're subjected monthly. Maybe that's why men declare war—because they have a need to bleed on a regular basis.

In Roz Warren
Glibquips (p. 107)

Crimmins, Cathy
A period is just the beginning of a lifelong sentence.

In Roz Warren
Glibquips (p. 107)

Unknown
What's the difference between worry and panic?

About twenty-eight days.

Source unknown

MICROBES

Donaldson, T.B.
He, who fights Microbes Away
Will be an Immune, some fine Day.

An Apropos Alphabet
Letter K

Dunne, Finley Peter
. . . mickrobes is a vigitable, an' ivry man is like a conservatory full iv
millyons iv these potted plants.

Mr. Dooley's Opinions
Christian Science (p. 5)

Gillilan, Strickland
Adam
Had 'em.

In Herbert V. Prochnow and Herbert V. Prochnow, Jr
A Treasury of Humorous Quotations

Huxley, Aldous
. . . think of the inexpugnable retreats for microbes prepared by
Michelangelo in the curls of Moses' beard!

Time Must Have a Stop
Chapter 3 (p. 36)

Osler, Sir William
In war the microbe kills more than the bullet.

In Harvey Cushing
The Life of Sir William Osler
Volume II (p. 427)

Unknown
In the nineteenth century men lost their fear of God and acquired a fear
of microbes.

Source unknown

Wolfe, Humbert

The doctor lives by chicken pox, by measles, and by mumps.
He keeps a microbe in a box and cheers him when he jumps.

Cursory Rhymes
Poems Against Doctors
II

God grant me . . . the insight to know the difference
Between a PMS day and a normal day . . .
Susan Hankla – (See p. 286)

MICROSCOPE

Dickinson, Emily
Faith is a fine invention
For gentlemen who see;
But microscopes are prudent
In an emergency.

Poems
Second Edition
XXX

Mayo, William J.
Modern medicine may be said to have begun with the microscope.

Journal of the Iowa Medical Society
Looking Backward and Forward in Medical Education
Volume 19, February 1929

NAUSEA

Siegel, Eli
Nausea can be unclearly accepted self-condemnation.

Damned Welcome
Aesthetic Realism Maxims
Part I, #124 (p. 37)

NERVE

Unknown

Phrenic nerve: C3, 4, 5 . . . keeps the diaphragm alive

Source unknown

NOUNS OF MULTITUDE

Unknown
A hive of allergists.
A body of anatomists.
A wiff of anesthesiologists.
A he(a)rd of audiologists.
A manipulation of chiropractors.
A clique of clinicians.
A wince of dentists.
A rash of dermatologists.
A guess of diagnosticians.
A crop of gastrologists.
A jury of generalists.
A smear of gynecologists.
A clot of hematologists.
A galaxy of hepatologists.
An obedience of hypnotists.
A harem of hysterologists.
An infusion of interns.
A practice of interns.
A scrub of interns.
A gargle of laryngologists.
A handful of mammologists.
A cuddle of nurses.
A giggle of nurses.
A help of nurses.
An eyeful of ophthalmologists.
A cast of orthopedic theumatologists.
A vision of orthopists.
A joint of osteopaths.
A body of pathologists.
A mixture of pharmacists.
A chest of phthisiologists.

An exercise of physical therapists.
A poll of pollenologists.
A pile of proctologists.
A couch of psychologists.
A train of pulmonologists.
A pew of rhinologists.
A column of spondylotherapists.
A cut of surgeons.
A flood of urologists.

Source unknown

NURSE

Anderson, Peggy
The nurse's job is to help the patients get well, or help them to die.

Nurse
Chapter 1 (p. 20)

. . . nurses play the same role on a regular floor that the electrocardiograph plays in the intensive care unit . . . They're the monitor.

Nurse
Chapter 1 (pp. 20–1)

Nurses do whatever doctors and janitors won't do.

Nurse
Chapter 2 (p. 31)

Barnes, Djuna
The only people who really *know* anything about medical science are the nurses, and they never tell; they'd get slapped if they did.

Nightwood
La Somnambule (p. 40)

Beckett, Samuel
The patients seeing so much of the nurses and so little of the doctor, it was natural that they should regard the former as their persecutors and the latter as their savior.

Murphy
Chapter 9 (p. 158)

Burns, Olive Ann
They ain't no feelin' in the world like takin' on somebody wilted and near bout gone, and you do what you can, and then all a-sudden the pore thang starts to put out new growth and git well.

Cold Sassy Tree
Chapter 3 (p. 12)

Chekhov, Anton
A doctor is called in, but a nurse sent for.

Note-Book of Anton Chekhov (p. 122)

Cowper, William
The nurse sleeps sweetly, hir'd to watch the sick,
Whom snoring she disturbs.

The Task
I
The Sofa, L. 89–90

Di Bacco, Babs Z.
Why modern doctors
Have more leisure time
For golf and cards
And things maritime
Than ever before
In history
While nurses don't,
Is a mystery.

American Journal of Nursing
Leisure Gap (p. 212)
January 1969

Euripides
Better be sick than tend the sick; the first is but a single ill, the last unites
mental grief with manual toil . . .

The Plays of Euripides
Hippolytus, L. 186

Hanson, Elayne Clipper
It seems to be a well known fact
That nurses fairly ooze with tact.
Their smiles so warm,
And full of charm,
Match voices, low
And movements slow.
Then why, if I may venture bold
Are nurses' hands so icy cold?

American Journal of Nursing
Paradox (p. 672)
March 1969

Jewett, Sarah Orne
She had no equal in sickness, and knew how to brew every old-fashioned
dose and to make every variety of herb-tea, and when her nursing was

put to an end by her patient's death, she was commander-in-chief at the funeral.

Deephaven
My Lady Brandon and the Widow Jim (p. 55)

Kipling, Rudyard

Let us now remember many honourable women,
Such as bade us turn again when we were like to die.

Collected Verse of Rudyard Kipling
Dirge of Dead Sisters

Lewis, Lucille

The central focus of nursing is to help the person cope with his physiological, psychological, and spiritual reactions to his health problems and maintain his integrity in these experiences. To be therapeutic, the nurse must contribute to the wholeness of man, to the interrelationships of the parts to the whole, to the person's here-and-now as well as to his future, to the health of all the parts so that the person may attain and maintain his highest potential. All of these are essential.

Nursing Outlook
This I Believe (p. 27)
Volume 15, Number 5, May 1968

Matthews, Marian

Where are the interns I recall?
Fountains of wisdom, one and all.
Men in white, mature and strong—
They were the Doctors, never wrong.

Something happened to them, or me;
They're not the giants they used to be.
With stethoscopes like shiny toys
They seem to me like little boys.

Excuse me for crying on your shoulder—
They're not younger—I am older.

American Journal of Nursing
Interns (p. 2492)
November 1968

Mayo, Charles H.

The trained nurse has given nursing the human, or shall we say, the divine touch, and made the hospital desirable for patients with serious ailments regardless of their home advantages.

Collected Papers of the Mayo Clinic & Mayo Foundation
The Trained Nurse
Volume 13, 1921

Movie
If you nurse as good as your sense of humor, I won't make it to Thursday.

The Sunshine Boys
Walter Mattheu to Rosettal LeNoire

Nightingale, Florence
A nurse who rustles (I am speaking of nurses professional and unprofessional) is the horror of a patient, though perhaps he does not know why.

Notes on Nursing
Noise (p. 47)

It seems a commonly received idea among men and even among women themselves that it requires nothing but a disappointment in love, the want of an object, a general disgust, or incapacity for other things to turn a woman into a good nurse.

Notes on Nursing
Conclusion (p. 133)

Never to allow a patient to be wakened, intentionally or accidentally, is a *sine qua non* of all good nursing.

Notes on Nursing
Noise (p. 44)

Paget, Stephen
"Talk of the patience of Job", said a Hospital nurse, *"Job was never on night duty."*

Confessio Medici
The Discipline of Practice (p. 83)

Ray, John
A *nurse's* tongue is privileged to talk.

A Complete Collection of English Proverbs (p. 17)

Richardson, Samuel
Male nurses are unnatural creatures!

Sir Charles Grandison
Part 2
Volume III
Letter XI (p. 58)

Roosevelt, Franklin Delano
. . . I urge that the Selective Service Act be amended to provide for the induction of nurses into the armed forces . . .

Annual Message to Congress
January 6, 1945

Schmitz, Jacqueline T.
Compressing an hour into a half,
Busy, yet heedful
Rushes the nurse
Bringing earnest solace to the
Sick and the needful.
Who must come first? How can she know?
Inverted scope, focused by Death,
Grants new perspective as He robs breath.

Journal of Nursing
Point of View (p. 2492)
November 1968

Stewart, Michael M.
When doctors doctor, and nurses nurse,
Most patients get better, though some get worse.
The system's not perfect, but one of the facts is
That no one is suing the nurse for malpractice:
She knows what her job is, and does it with grace,
While doctors make sure that she stays in her place.

The New England Journal of Medicine
Help? (p. 1384)
Volume 285, Number 24, 1971

Unknown
[Nurse] A panhandler.

Esar's Comic Dictionary

To be a good nurse you must be absolutely sterile.

In Alexander Abingdon
Bigger & Better Boners (p. 74)

NURSE'S PRAYER

Hargrove, Cecilia
I entertain no great ambition
For lots of dough or high position.
One thought propels me through the night—
I hope my I.V.'s come out right!

American Journal of Nursing
Prayer of a Pediatric Night Nurse (p. 696)
March 1968

Unknown
Dear Lord, please give me strength,
To face the day ahead.
Dear Lord, please give me courage,
As I approach each hurting bed.
Dear Lord, please give me wisdom
With every word I speak.
Dear Lord, please give me patience,
As I comfort the sick and weak.
Dear Lord, please give me assurance,
As the day slips into night.
That I have done the best I can,
That I have done what's right.

Source unknown

As I care for my patients today
Be there with me, Oh Lord, I pray.
Make my words kind—it means so much.
And in my hands place your healing touch.
Let your love shine through all that I do,
So those in need may hear and feel You.

Source unknown

Let me dedicate my life today, in the care of those who come my way.
Let me touch each one with healing hand, and the gentle art for which
 I stand.
And then tonight when day is done, let me rest in peace, if I've helped
 just one.

Source unknown

Lord, give me strength, compassion, understanding, skill and tenderness
for this selfless service.

Source unknown

I'm a Dedicated Nurse,
Lord I pray I do my part
to lift each downcast spirit
and to soothe
each heavy heart.
May my touch bring reassurance.
May my voice be soothing too.
May my gentle care remind them.
Of the love they have in You.

Source unknown

Give to my heart, Lord, compassion and understanding. Give to my
hands, skill and tenderness. Give to my ears the ability to listen. Give
to my lips words of comfort. Give to me, Lord, strength for this selfless
service and enable me to give hope to those I am called to serve.

Source unknown

NUTRITION

Baum, Harold
If you gobble tagliatelli,
Chicken soup with vermicelli,
You'll acquire a sagging belly—
What's the use of that?
If your intake's calorific,
Guzzling beer till soporific,
Possibly you'll feel terrific,
But you'll end up fat.
Fat against starvation; fat for insulation;
If you sit hard you'll bounce on lard
Which substitutes in females for inflation.
Fat provides when you are needing
Glucogenic when you're seeding
Product of excessive feeding—
Hail adipocyte!

The Biochemists' Handbook
Fatty Acid Biosynthesis
(Sung To The Tune Of "Men of Harlech")

Davis, Adelle
Nutrition is a young subject; it has long been kicked around like a puppy that cannot take care of itself. Food faddists and crackpots have kicked it pretty cruelly . . . They seem to believe that unless food tastes like Socratic hemlock, it cannot build health. Frankly, I often wonder what such persons plan to do with good health in case they acquire it.

Let's Eat Right to Keep Fit
Chapter 1 (p. 3)

. . . eat breakfast like a king, lunch like a prince, and dinner like a pauper.

Let's Eat Right to Keep Fit
Chapter 2 (p. 19)

Unknown

To help heal wounds, we're told to drink Zinc.
Comes next, perhaps to help the skin, Tin.
Our poor, tired blood's long been requirin' Iron,
And soon we'll need, if health's not proper, Copper,
For stomach upset and for gas, Brass,
To make us gleam abroad, at home, Chrome,
And for that cough and throaty tickle, Nickel.
And what's the best of cures? I'm told Gold.

Quote, The Weekly Digest
May 7, 1967 (p. 378)

When taken,
To be well shaken.
George Colman (the Younger) – (See p. 205)

OBSERVATION

George, W.H.
Not a single detail escapes the glance of the lynx-eyed detective of fiction, but his like has never been found alive. It seems then that if the eye is used to observe it may be necessary not only to look *at* the detail to be observed, but also to look *for* it. We do not always find with our sense organs unless we also seek.

The Scientist in Action. A Scientific Study of his Methods
The Eye-Witness's Observation (p. 84)

Nightingale, Florence
In dwelling upon the vital importance of *sound* observation, it must never be lost sight of what observation is for. It is not for the sake of piling up miscellaneous information or curious facts, but for the sake of saving life and increasing health and comfort.

Notes on Nursing
Observation of the Sick (p. 125)

OINTMENT

Carroll, Lewis

"In my youth," said the sage, as he shook his grey locks,
"I kept all my limbs very supple
By the use of this ointment—one shilling the box—
    Allow me to sell you a couple."

The Complete Works of Lewis Carroll
Alice's Adventures in Wonderland
Chapter 5

Ecclesiastes 10:1

Dead flies cause the ointment of the apothecary to send forth a stinking savor.

The Bible

OPERATION

Shaw, George Bernard
The epithet *beautiful* is used by surgeons to describe operations which the patient describes as ghastly.

In Evan Esar
20,000 Quips and Quotes

Now there is no calculation that an engineer can make as to the behavior of a girder under a strain, of an astronomer as to the recurrence of a comet, more certain than the calculation that under such circumstances we shall be dismembered unnecessarily in all directions by surgeons who believe the operations to be necessary solely because they want to perform them.

The Doctor's Dilemma
Preface On Doctors (p. vi)

Unknown
A minor operation is one performed on someone else.

Source unknown

Recovering from an operation, a patient asked the doctor why all the blinds were drawn. The doctor replied, "Well, you see, there's a fire across the street and I didn't want you to wake up thinking the operation had been a failure."

Source unknown

OPINIONS

Holmes, Oliver Wendell
It is not often that an opinion is worth expressing, which cannot take care of itself.

Medical Essays
Border Lines in Medical Science (pp. 271–2)

Lippmann, Walter
True opinions can prevail only if the facts to which they refer are known; if they are not known, false ideas are just as effective as true ones, if not a little more effective.

Liberty and the News (p. 71)

Locke, John
New opinions are always suspected, and usually opposed, without any other reason but because they are not already common.

An Essay Concerning Human Understanding
Dedicatory Epistle

Unknown
. . . opinions ought to count by weight rather than by number . . .

In James Joseph Sylvester
Collected Mathematical Works
Volume III
Address on Commemoration Day at Johns Hopkins University (p. 73)

ORAL HYGIENE

Kraus, Jack
Oral hygiene—Strict gum control.

Quote, The Weekly Digest
August 4, 1968 (p. 97)

PAIN

Coates, Florence Earle
Ah, me! the Prison House of Pain!—what lessons
   there are bought!—
Lessons of a sublimer strain
Than any elsewhere taught.

Poems
Volume Two
The House of Pain

Dickinson, Emily
Pain has an element of blank;
It cannot recollect
Where it began, or if there were
A day when it was not.

Poems (1890–1896)
XIX, The Misery of Pain

Emerson, Ralph Waldo
He has seen but half the universe who never has been shewn the House
of Pain.

Natural History of Intellect
The Tragic (p. 260)

Howell, James
Pains is the price, that God puttuth upon all things.

Proverbs (p. 19)

Johnson, Samuel
. . . those who do not feel Pain, seldom think that it is felt . . .

The Rambler
Number 48 (p. 284)

Latham, Peter Mere
It would be a great thing to understand Pain in all its meanings.

In William B. Bean
Aphorisms from Latham (p. 71)

Preston, Margaret Junkin
Pain is no longer pain when it is past . . .

Old Songs and New
Nature's Lesson (p. 260)

Schweitzer, Albert
Whosoever is spared personal pain must feel himself called to help in diminishing the pain of others.

Recalled on his death
September 4, 1965

Thompson, Francis
Nothing begins, and nothing ends,
   That is not paid with a moan;
For we are born in other's pain,
   And perish in our own.

Complete Poetical Work of Francis Thompson
Daisy
Stanza 15

Watson, William
Pain with the thousand teeth.

The Poems of William Watson
The Dream of Man (p. 127)

PARASITE

Bishop of Birmingham
The loathsome parasite is a result of the integration of mutations; it is both an exquisite example of adaptation to environment and ethically revolting.

Nature
November 29, 1930

Wilson, Edward O.
Leishmaniasis, schistosomiasis, malignant tertian malaria, filariasis, echinococcosis, onchocerciasis, yellow fever, amoebic dysentery, bleeding bot-fly cysts . . . evolution has devised a hundred ways to macerate livers and turn blood into a parasite's broth.

Biophilia
Bernhardsdorp (pp. 12–3)

PATIENTS

Armour, Richard
The perfect patient let us praise:
He's never sick on Saturdays,
In fact this wondrous, welcome sight
Is also never sick at night.
In waiting rooms he does not burn
But gladly sits and waits his turn,
And even, I have heard it said,
Begs others, "Please go on ahead."
He takes advice, he does as told,
He has a heart of solid gold.
He pays his bills, without a fail,
In cash, or by the same day's mail.
He has but one small fault I'd list:
He doesn't (what a shame!) exist.

The Medical Muse
Ideal Patient

Bahya, ben Joseph ibn Paauda
For a sick person who lies to his physician cheats only himself, wastes the physician's efforts and aggravates his sickness.

Duties of the Heart
Third Treatise
Chapter V (p. 56)

Cushing, Harvey
Every patient, he said, provided two questions—firstly what can be learnt from him and secondly what can be done for him.

In Robert Coope
The Quiet Art (p. 103)

241

de Madariaga, Salvadore
There are no diseases, there are only patients.

Essays with a Purpose
On Medicine (p. 174)

Drake, Daniel
[There is no era in the life of a physician] in which his self-complacency is
so exalted, as the time which passes between receiving his diploma with
its blue ribbon, and receiving crepe and gloves, to wear at the funeral of
his first patient.

Western Journal of Medicine and Surgery
N.S. II:355, October 1844

Dunne, Finley Peter
A patient in th' hands iv a doctor is like a hero in th' hands iv a story
writer. He's goin' to suffer a good dale, but he's goin' to come out all
right in th' end.

Mr. Dooley: On Making a Will and Other Necessary Evils
Going to See the Doctor

Eliot, T.S.
Reilly: Most of my patients begin, Miss Coplestone,
By telling me exactly what is the matter with them.
And what I am to do about it.

The Cocktail Party
Act 2 (p. 131)

Helmuth, William Tod
She sent for me in haste to come and see,
What her condition for a cure might be.
Dear me! a patient—what a happy tone,
To have a patient and one all my own—
To have a patient and myself be feed,
Raised expectations very high indeed—
I saw a practice growing from the seed.

Scratches of a Surgeon
My First Patient (p. 61)

Heschel, Abraham J.
The patient must not be defined as a client who contracts a physician for
service; he is a human being entrusted to the cure of a physician.

The Insecurity of Freedom
The Patient as a Person (p. 31)

Holmes, Oliver Wendell

What I call a good patient is one who, having found a good physician, sticks to him till he dies.

Medical Essays
The Young Practitioner (p. 390)

Once in a while you will have a patient of sense, born with the gift of observation, from whom you may learn something.

Medical Essays
The Young Practitioner (pp. 382–3)

If you are making choice of a physician, be sure to get one, if possible, with a cheerful and serene countenance.

The Professor at the Breakfast Table (p. 159)

Hubbard, Kin

"Be kind t' th' henn egg. When sickness enters th' home an' th' patient comes thru th' crisis twenty pounds lighter than a straw hat, an' is propped up with pillows in th' bay window t' watch th' speedin', an' loved ones try t' tempt him with round steak, an' pickles an' near beer, he wearily waves 'em away. But with his first returnin' strength he squirms an' turns his listerless eyes toward th' kitchen an' says, in a voice weak an' scarcely audible, 'Maw, I believe I could worry down an egg,'". . .

Abe Martin: Hoss Sense and Nonsense (p. 53)

Mayo, William J.

. . . the highly scientific development of this mechanistic age had led perhaps to some loss in appreciation of the individuality of the patient and to trusting largely to the laboratories and outside agencies which tended to make the patient not the hub of the wheel, but a spoke.

Collected Papers of the Mayo Clinic & Mayo Foundation
Edward Martin, M.D., 1859–1938
Volume 30, 1938

Morris, Robert Tuttle

It is the patient rather than the case which requires treatment.

Doctors versus Folks
Chapter 2

Newman, Sir George

There are four questions which in some form or other every patient asks his doctor:

(a) What is the matter with me? This is *diagnosis*.
(b) Can you put me right? This is *treatment and prognosis*.

(c) How did I get it? This is *causation*.
(d) How can I avoid it in future? This is *prevention*.

The Lancet
Preventive Medicine for the Medical Student (p. 113)
Volume 221, November 21, 1931

Osler, Sir William
To study the phenomena of disease without books is to sail an uncharted sea, while to study books without patients is not to go to sea at all.

In Harvey Cushing
The Life of Sir William Osler
Volume I (p. 67)

Parrot, Max
It has often been said that the technical aspects of medicine are easy. The difficult part is dealing with the personality of the patient, the so-called psychological or human factor. This takes up a great deal of the time of the practicing physician. It is harder on the doctor's constitution than all of the technical aspects of medicine. It may even cause his or her demise, in the case of a physician with an autonomic nervous system that can't take the heat.

In Irving Oyle
The New American Medical Show (p. 25)

Potter, Stephen
If Patient turns out to be really ill, it is always possible to look grave at the same time and say 'You realise, I suppose, that 25 years ago you'd have been dead?'

One-Upmanship
Chapter II (p. 28)

Rhazes
The patient who consults a great many physicians is likely to have a very confused state of mind.

In Samuel Evans Massengill
A Sketch of Medicine and Pharmacy (p. 45)

Sacks, Oliver
There is only one cardinal rule: one must always *listen* to the patient.

Newsweek
Listening to the Lost (p. 70)
August 20, 1984

Shakespeare, William
MACBETH: How does your patient, doctor?
DOCTOR: Not so sick, my lord,
             As she is troubled with thick-coming fancies
             That keep her from her rest.

Macbeth
Act V, Scene III, L. 37–40

Unknown
Quoth Doctor Squill of Ponder's End,
"Of all the patients I attend
  Whate'er their aches or ills,
None ever will my fame attack."
"None ever can," retorted Jack,
  "For dead men tell no tales."

In William Davenport Adams
English Epigrams
The Doctor's Security
cclxxxix

The patient is never out of danger as long as the doctor continues to make visits.

In Evan Esar
20,000 Quips and Quotes

Practicing medicine would be a great life if it weren't for patients.

Source unknown

PHARMACIST

Eisenschiml, Otto

The surgeon who has performed scores of brilliant operations is less talked about than the one who has inadvertently killed a patient; the pharmacist who has carefully filled prescriptions for a lifetime remains obscure, but will gain publicity by a single oversight.

The Art of Worldly Wisdom
Part Seven (p. 87)

Flexner, Abraham

The physician thinks, decides, and orders; the pharmacist obeys—obeys, of course, with discretion, intelligence, and skill—yet, in the end, obeys and does not originate. Pharmacy therefore is an arm added to the medical profession, a specially and distinctly higher form of handicraft, not a profession . . .

School and Society
Is Social Work a Profession?
Volume 1, 1915 (p. 905)

Unknown

A piller of the community.

Esar's Comic Dictionary

PHARMACY

Ghalioungui, Paul
The word *pharmakon*, whence pharmacy is derived, meant in Greek not only medicament, poison, or magical procedure, but also that which is slain to expiate the crimes of a city, like the scapegoat of Biblical times . . . In other words, it meant 'what carries off disease'.

Magic and Medical Science in Ancient Egypt
Chapter II (p. 35)

PHYSIC

Colton, Charles Caleb

No men despise physic so much as physicians, because no men so thoroughly understand how little it can perform.

Lacon
1:179

Heurnius

Many of them to get a fee, will give physic to every one that comes, when there is no cause.

In William Tod Helmuth
Scratches of a Surgeon (p. 9)

Holmes, Oliver Wendell

Not to take authority when I can have facts; not to guess when I can know; not to think a man must take physic because he is sick.

In Robert Coope
The Quiet Art (p. 101)

Lettsom, J.C.

When people's ill, they comes to I,
  I physics, bleed, and sweats 'em;
Sometimes they live, sometimes they die.
  What's that to I? I lets 'em.

In William Davenport Adams
English Epigrams
On Dr. Lettsom, by Himself
cclxxii

Milton, John

. . . in Physic things of melancholic hue and quality are us'd against melancholy, sour against sour, salt to remove salt humours.

Samson Agonistes
On that sort of Dramatic Poem which is call'd Tragedy (p. 79)

Pope, Alexander
Learn from the beasts the physic of the field.

The Poems of Alexander Pope
Volume III
Essay on Man
Epis. iii, L. 174

Proverb
Dear physic always does good, if not to the patient at least to the apothecary.

Source unknown

Warre and Physicke are governed by the eye.

In George Herbert
Outlandish Proverbs
#906

Ray, John
If physic do not work, prepare for the kirk.

A Complete Collection of English Proverbs (p. 149)

Shakespeare, William
Take physic, pomp;
Expose thyself to feel what wretches feel.

King Lear
Act III, Scene II, L. 33–34

'Tis time to give 'em physic, their diseases
Are grown so catching.

King Henry the Eighth
Act I, Scene III, L. 36–37

In this point
All his tricks founder, and he brings his physic
After his patient's death.

King Henry the Eighth
Act III, Scene II, L. 39–41

Throw physic to the dogs; I'll none of it.

Macbeth
Act V, Scene III, L. 47

Unknown

Dear physic always does good, if not to the patient at least to the apothecary.

Source unknown

I was well, would be better, took physic and died.

Source unknown

Physic brings health, and Law promotion,
  To followers able, apt, and pliant;
But very seldom, I've a notion,
  Either to patient or to client.

In William Davenport Adams
English Epigrams
On Law and Physic
dcccxliv

No good physician ever takes physic.
Unknown – (See p. 278)

PHYSICAL

Chesterfield, Philip Dormer Stanhope
Physical ills are the taxes laid upon this wretched life; some are taxed higher, and some lower, but all pay something.

The Letters of Philip Dormer Stanhope
Volume Five
#2031, 22 November, 1757 (p. 2265)

Livermore, Mary
Other books have been written by men physicians . . . One would suppose in reading them that women possess but one class of physical organs, and that these are always diseased. Such teaching is pestiferous, and tends to cause and perpetuate the very evils it professes to remedy.

What Shall We Do with Our Daughters?
Chapter II (p. 22)

PHYSICIAN

Alexander the Great
I am dying with the help of too many physicians.

Attributed

Allman, David
The dedicated physician is constantly striving for a balance between personal, human values, scientific realities and the inevitabilities of God's will.

Address to National Conference of Christians and Jews
The Brotherhood of Healing
February 12, 1958

Aristotle
It is the physician's business to know that circular wounds heal more slowly, the geometer's to know the reason why.

Posterior Analytics
79a14

A patient should call in a physician; he will not get better if he is doctored out of a book.

Politics
1287a33

Aurelius, Marcus [Antoninus]
Think continually how many physicians are dead after often contracting their eyebrows over the sick . . .

Meditations
Book IV, Section 48

Bacon, Francis
Physicians are some of them so pleasing and comfortable to the humor of the patient, as they press not the true cure of the disease; and some other are so regular in proceeding according to art for the disease, as

they respect not sufficiently the condition of the patient. Take one of a middle temper; or if it may not be found in one man, combine two of either sort; and forget not to call as well the best acquainted with your body, as the best reputed of for his faculty . . .

> Essays, Advancement of Learning, New Atlantis, and Other Pieces
> The Essayes or Counsels, Civil and Morall
> Of Regiment of Health (p. 94)

. . . for the weakness of patients, and the sweetness of life, and nature of hope, maketh men depend upon physicians with all their defects.

> Advancement of Learning
> Second Book, X, 2

. . . it is the office of a physician not only to restore health, but to mitigate pain and dolors; and not only when such mitigation may conduce to recovery, but when it may serve to make a fair and easy passage.

> Advancement of Learning
> Second Book, X, 7

Bailey, Percival
The function of the physician is to cure a few, help many, and comfort all.

> Perspectives
> Volume 4, Number 254, 1961

Baldwin, Joseph G.
Nobody knew who or what they were, except as they claimed, or as a surface view of their characters indicated. Instead of taking to the highway and magnanimously calling upon the wayfarer to stand and deliver, or to the fashionable larceny of credit without prospect or design of paying, some unscrupulous horse doctor would set up his sign as "Physician and Surgeon" and draw his lancet on you, or fire at random a box of pills into your bowels, with a vague chance of hitting some disease unknown to him, but with a better prospect of killing the patient, whom or whose administrator he charged some ten dollars a trial for his marksmanship.

> The Flush Times of Alabama and Mississippi
> How the Times Served the Virginians (p. 89)

Bass, Murray H.
The ideal physician should be a combination of three persons—a clergyman, a fireman and a scientist. He must know how to handle and console the patient and his family . . . he must be ready to answer an "alarm" day and night; he must know the science of medicine . . . using its present potentialities to the utmost of his ability.

> Clinical Pediatrics
> Volume 3, Number 50, 1964

Bernard, Claude

Most physicians seem to believe that, in medicine, laws are elastic and indefinite. These are false ideas which must disappear if we mean to found a scientific medicine. As a science, medicine necessarily has definite and precise laws which, like those of all the sciences, are derived from the criterion of experiment.

Experimental Medicine
II, 2

Bierce, Ambrose

PHYSICIAN, *n*. One upon whom we set our hopes when ill and our dogs when well.

The Enlarged Devil's Dictionary

Bigelow, Jacob

I would answer that *he is a great physician who, above other men, understands diagnosis*. It is not he who promises to cure all maladies, who has a remedy ready for every symptom, or one remedy for all symptoms; who boasts that success never fails him, when his daily history gives the lie to such assertion. It is rather he, who, with just discrimination, looks at a case in all its difficulties; who to habits of correct reasoning, adds the acquirements obtained from study and observation; who is trustworthy in common things for his common sense, and in professional things for his judgment, learning and experience; who forms his opinion positive or approximative, according to the evidence; who looks at the necessary results of inevitable causes; who promptly does what man may do of good, and carefully avoids what he may do of evil.

Nature in Disease
1852

Bonaparte, Napoleon

In my opinion physicians kill as many people as we generals.

In J. Christopher Herold (Editor)
The Mind of Napoleon
Science and the Arts (p. 137)

You are a physician, doctor. You would promise life to a corpse if he could swallow pills . . .

In J. Christopher Herold (Editor)
The Mind of Napoleon
Science and the Arts (pp. 137–8)

Boorde, Andrew

A good cook is half a physician. For the chief physic (the counsel of a physician excepted) doth come from the kitchen; wherefore the physician and the cook for sick men must consult together for the preparation of meat for sick men. For if the physician, without the cook, prepared any meat, except he be very expert, he will make a wearish dish of meat, the which the sick cannot take.

The Wisdom of Andrew Boorde (p. 49)

Brackenridge, Hugh Henry
Gravity is the most practical qualification of the physician.

Modern Chivalry
Part II, Volume I, Chapter X (p. 378)

Brown, John
The prime qualifications of a physician may be summed up in the words *Capax, Perspicax, Sagax, Efficax, Capax*—there must be room to receive, and arrange and keep knowledge; *Perspicax*—senses and perceptions keen, accurate, and immediate, to bring in materials from all sensible things; *Sagax*—a central power of worth, of choosing and rejecting, of judging; and finally, *Efficax*—the will and the way—the power to turn all the other three—capacity, perspicacity, sagacity, to account, in the performance of the thing in hand, and thus rendering back to the outer world, in a new and useful form, what you have received from it.

Horae Subsecivae
Series II
With Brains Sir

Buchan, William
No two characters can be more different than that of the honest physician and the quack; yet they have generally been much confounded.

Domestic Medicine
Introduction (p. xvi)

Physicians, like other people, must live by their employment.

Domestic Medicine
Introduction (p. xviii)

Burgess, Anthony
Keep away from physicians. It is all probing and guessing and pretending with them. They leave it to Nature to cure in her own time, but they take the credit. As well as very fat fees.

Nothing Like the Sun
Chapter VIII (p. 180)

Byron, Lord George Gordon
This is the way physicians mend or end us,
Secundum artem: but although we sneer
In health—when ill, we call them to attend us,
Without the least propensity to jeer.

Don Juan
Canto X, Stanza XLII

Cambyses

. . . Physicians were like Taylers and Coblers, the one mended our sicke bodies, as the other did our cloaths.

In Robert Burton
The Anatomy of Melancholy
Volume II
Part 2, Section 41 (p. 211)

Camden, William

Few physicians live well.

Remaines Concerning Britain
Proverbs (p. 322)

Carlyle, Thomas

The healthy know not of their health, but only the sick: this is the physician's aphorism.

Characteristics (p. 3)

Chronicles II 16:12–13

Asa . . . was diseased in his feet, until his disease was exceedingly great: yet in his disease he sought not to the Lord, but to the physicians.

The Bible

Clowes, William

When a physician or a surgeon comes to a man that lies sick and is in danger of death, yet by his judgment and skill, promises with God's help to cure him of his griefs and maladies, then the sick patient greatly rejoices and presently compares him to a god. But afterwards, being somewhat recovered, and perceiving good amendment, he says he is but an angel and not a god. Again, after he begins to walk abroad and to fall to his meat, truly he is then accounted no better than a man. In the end, when he happily comes for his money for the curing of his grievous sickness, he now reports him to be a devil and shuts the door.

Selected Writings of William Clowes
A Tragical History (p. 63)

Collins, Joseph

The longer I practice medicine, the more I am convinced every physician should cultivate lying as a fine art. There are lies which contribute enormously to the success of the physician's mission of mercy and salvation.

Quoted in *Reader's Digest*
May 1933

Colton, Charles Caleb

Physicians must discover the weaknesses of the human mind, and even condescend to humor them, or they will never be called in to cure the infirmities of the body.

Lacon
1.482

Croll, Oswald

A Physition should be born out of the Light of Grace and Nature of the inward and invisible Man . . .

Philosophy Reformed and Improved in Four Profound Tractates (p. 22)

. . . a Physitian therefore should have both the Theory and Practice, he must both know and prepare his medicines . . .

Philosophy Reformed and Improved in Four Profound Tractates (p. 152)

Cushing, Harvey

A physician is obligated to consider more than a diseased organ, more even than the whole man—he must view the man in his world.

In René Dubos
Man Adapting
Chapter XII (p. 342)

Davies, Robertson

I delivered my body into the hands of Learned Physicians this morning confiding that they may discover why I have hay fever. As soon as they got me out of my clothes I ceased to be a man to them, and they began to talk about me as though I did not understand English.

The Table Talk of Samuel Marchbanks (p. 194)

de Belleville, Nicholas

If you get one *good* doctor, you get one *good* thing, but if you get one *bad* doctor, you get one *bad* thing. If you have a lawsuit, you get a bad lawyer, you lose your suit—you can appeal; but if you have one bad doctor, and he kills you, then there can be no appeal.

In Stephen Wickes
History of Medicine in New Jersey
Part 2

Deeker, Thomas

A good physician comes to thee in the shape of an angel, and therefore let him boldly take thee by the hand, for he has been in God's garden, gathering herbs and sovereign roots to cure thee. The good physician deals in simples and will be simply honest with thee in they preservation.

In Robert Coope
The Quiet Art (p. 192)

Donne, John
I observe the *Physician* with the same diligence, as hee the *disease*.

Devotions Upon Emergent Occasions
Meditation
VI (p. 29)

Whilst my Physitians by their love are growne
Cosmographers, and I their Mapp, who lie
Flat on this bed, that by them may be showne
That this is my South-west discoverie
Per fretum febris, by these streights to die.

The Poems of John Donne
Hymne to God My God, in My Sicknesse, L. 6–10

Drake, Daniel
The young physician is not aware how soon his elementary knowledge—much of which is historical and descriptive, rather than philosophical—will fade from his mind, when he ceases to study. That which he possesses can only be retained by new additions.

Practical Essays on Medical Education, and the Medical Profession (p. 61)

Professional fame, is the capital of a physician, and he must not suffer it to be purloined, even should its defence involve him in quarrels.

Practical Essays on Medical Education, and the Medical Profession (p. 99)

Dryden, John
The first Physicians by Debauch were made:
Excess began, and Sloth sustains the Trade.

The Poems of John Dryden
Volume IV
To John Dryden, of Chesterton
L. 73 (p. 1530)

Duffy, John C.
Litin, Edward M.
These are the duties of a physician: First . . . to heal his mind and to give help to himself before giving it to anyone else.

Journal of the American Medical Association
Psychiatric Morbidity of Physicians (p. 989)
Volume 189, 1964

Ecclesiasticus 38:1
Honour a physician with the honour due unto him for the uses which
you may have of him: for the Lord hath created him.

The Bible

Ecclesiasticus 38:15
He that sinneth before his Maker, let him fall into the hands of a
physician.

The Bible

Eisenschiml, Otto
The wise physician . . . knows when not to prescribe . . .

The Art of Worldly Wisdom
Part Six (p. 71)

The wise physician avoids the knife; if he prescribes a bitter draft, he
prescribes it in small doses or sweetens it to disguise its taste.

The Art of Worldly Wisdom
Part Seven (p. 87)

Emerson, Ralph Waldo
The physician prescribes hesitatingly out of his few resources . . . If the
patient mends, he is glad and surprised.

Complete Works
Volume 6
Considerations by the Way (p. 245)

Field, Eugene
When one's all right, he's prone to spite
    The doctor's peaceful mission;
But when he's sick, it's loud and quick
    He bawls for a physician.

The Poems of Eugene Field
Doctors

No matter what conditions
    Dyspeptic come to feaze,
The best of all physicians
    Is apple pie and cheese!

The Poems of Eugene Field
Apple Pie and Cheese
Stanza 5

Fielding, Henry
Every physician almost hath his favorite disease.

Tom Jones
Book II, Chapter 9 (p. 33)

The gentleman of the Æsculapin art are in the right in advising, that the moment the disease has entered at one door, the physician should be introduced at the other.

Tom Jones
Book V, Chapter 7 (p. 86)

. . . as a wise general never despises his enemy, however inferior that enemy's force may be, so neither doth a wise physician ever despise a distemper, however inconsiderable.

Tom Jones
Book V, Chapter 7 (p. 90)

Florio, John
From the phisito & Attorney,
keepe not the truth hidden.

Firste Fruites
Proverbs, Chapter 19

Unto a deadly disease, neyther
Phisition nor phisick wil serve.

Firste Fruites
Proverbs, Chapter 19

Ford, John
Physicians are the bodies coblers, rather than the Botchers, of men's bodies; as the one patches our tattered clothes, so the other solders our diseased flesh.

The Lovers Melancholy
Act I, Scene I (p. 13)

Fox, Sir Theodore
The patient may well be safer with a physician who is naturally wise than with one who is artificially learned.

The Lancet
Purposes of Medicine
Volume 2, October 23, 1965 (p. 801)

Fuller, Thomas
Commonly Physicians like beer are best when they are old; & Lawyers like bread when they are young and new.

The Holy State
The Second Book, Chapter 1 (p. 52)

Every man is a fool or a physician at forty.

Gnomologia
Number 1428

Gisborne, Thomas

It is frequently of much importance, not to the comfort only, but to the recovery of the patient, that he should be enabled to look upon his Physician as his friend.

An Enquiry into the Duties of Men
The Duties of Physicians (p. 398)

Gracián, Baltasar

The wise physician, if he has failed to cure, looks out for someone who, under the name of consultation, may help him carry out the corpse.

In Herbert V. Prochnow and Herbert V. Prochnow, Jr
A Treasury of Humorous Quotations

Hazlitt, William Carew

One said a Physitian was naturall brother to the wormes, because he was ingendered out of mans corruption.

Shakespeare Jest Books
Volume III
Conceit, Clichés, Flashes and Whimzies
Number 42

One said Physitians had the best of it; for, if they did well, the world did proclaime it; if ill, the earth did cover it.

Shakespeare Jest Books
Volume III
Conceit, Clichés, Flashes and Whimzies
Number 127

Heraclitus

The physicians . . . cutting, cauterizing, and in every way torturing the sick, complain that the patients do not pay them fitting reward for thus effecting these benefits and sufferings.

On Nature
Fragments, LVIII (p. 98)

Herophilus

He is the best physician who knows how to distinguish the possible from the impossible.

In Samuel Evans Massengill
A Sketch of Medicine and Pharmacy (p. 28)

Heschel, Abraham J.

What manner of man is the doctor? Life abounds in works of achievement, in areas of excellence and beauty, but the physician is a person who has chosen to go to the areas of distress, to pay attention to sickness and affliction, to injury and anguish.

The Insecurity of Freedom
The Patient as a Person (p. 28)

Hippocrates
. . . physicians are many in title but very few in reality.

Laws
I (p. 144)

Hoffmann, Friedrich
The physician is the servant of nature, not her master; the principles of nature and of art are the same and hence the physician must work and act with nature.

Fundamenta Medicianae
Physiology
Chapter I, 6 (p. 5)

The perfect physician must have not only the knowledge of medical art but also prudence and wisdom.

Fundamenta Medicianae
Physiology
Chapter I, 10 (p. 6)

Holmes, Oliver Wendell
. . . a physician's business is to avert disease, to heal the sick, to prolong life, and to diminish suffering . . .

Medical Essays
Scholastic and Bedside Teaching (p. 274)

A man of very moderate ability may be a good physician if he devotes himself faithfully to the work.

Medical Essays
Scholastic and Bedside Teaching (p. 300)

But the practising physician's office is to draw the healing waters, and while he gives his time to this labor he can hardly be expected to explore all the sources that spread themselves over the wide domain of science. The traveler who would not drink of the Nile until he had tracked it to its parent lakes, would be like to die of thirst; and the medical practitioner who would not use the results of many laborers in other departments without sharing their special toils, would find life far too short and art immeasurably too long.

Medical Essays
Scholastic and Bedside Teaching (p. 274)

The face of a physician, like that of a diplomat, should be impenetrable.

Medical Essays
The Young Practitioner (p. 388)

A physician who talks about *ceremony* and *gratitude*, and *services rendered*, and the *treatment he got*, surely forgets himself . . .

Medical Essays
The Contagious of Puerperal Fever (p. 115)

The life of a physician becomes ignoble when he suffers himself to feed on petty jealousies and sours his temper in perpetual quarrels.

Medical Essays
The Young Practitioner (p. 392)

. . . the old age of a physician is one of the happiest periods of his life. He is loved and cherished for what he has been, and even in the decline of his faculties there are occasions when his experience is still appealed to, and his trembling hands are looked to with renewing hope and trust . . .

Medical Essays
The Young Practitioner (p. 395)

Howells, William Dean
I do not know how it is that clergymen and physicians keep from telling their wives the secrets confided to them; perhaps they can trust their wives to find them out for themselves whenever they wish.

The Rise of Silas Lapham
Chapter 27 (p. 360)

Hubbard, Elbert
DOCTOR: 1. A person who has taken seriously the biblical injunction, "Physician, heel thyself!"

The Roycroft Dictionary (p. 40)

Hufeland, Christoph Wilhelm
The physician must generalize the disease, and individualize the patient.

In Oliver Wendell Holmes
Medical Essays
Scholastic and Bedside Teaching (p. 275)

Jackson, James
I have often remarked that, though a physician is sometimes blamed very unjustly, it is quite as common for him to get more credit than he is fairly entitled to; so that he has not, on the whole, any right to complain.

Letters to a Young Physician
1885

Jekyll, Joseph
See, one physician, like a sculler, plies,
The patient lingers and by inches dies.
But two physicians, like a pair of oars,
Waft him more swiftly to the Stygian shores.

In Herbert V. Prochnow and Herbert V. Prochnow, Jr
A Treasury of Humorous Quotations

Jeremiah 8:22
Is there no balm in Gilead: is there no physician there?

The Bible

John of Salsbury
The common people say, that physicians are the class of people who kill
other men in the most polite and courteous manner.

Polycraticus
Book II, Chapter 29

Johnson, Samuel
A physician in a great city seems to be the mere plaything of fortune; his
degree of reputation is for the most part totally casual; they that employ
him know not his excellence; they that reject him know not his deficience.

In William Osler
Aequanimitas (p. 132)

Jonsen, Albert R.
. . . the absolute asceticism of the residency recreates, for the young
physician, the sacrificial ethic of monastic medicine. That ethic is service:
immediate response to the emergency room, to the demands of reports,
unmitigated responsibility for correct decisions made promptly and
communicated clearly; flagellating denial of sleep, self-indulgence, and
frivolity, even to the point of depression and deterioration of personal
life, of friendship and love.

New England Journal of Medicine
Watching the Doctor
Volume 308, Number 25, June 23, 1983 (p. 1534)

La Bruyère, Jean
As long as men are liable to die and desirous to live, a physician will be
made fun of, but he will be well paid.

Characters
14.65

Lamb, William
English physicians kill you, the French let you die.

In Elizabeth Longford
Queen Victoria
Chapter 5 (p. 69)

Latham, Peter Mere
I am persuaded that when the physician is called upon to perform great things, even to arrest destructive disease, and to save life, his skill in wielding the implements of his art rests mainly upon the right understanding of simple and single indications, and of the remedies which have power to fulfill them.

In William B. Bean
Aphorisms from Latham (p. 19)

There are always two parties of the management of the disease—the physician and the patient.

In William B. Bean
Aphorisms from Latham (p. 21)

But Nature, in all her powers and operations, allows herself to be led, directed, and controlled. And to lead, direct, or control for purposes of good, this is the business of the physician.

In William B. Bean
Aphorisms from Latham (p. 24)

The best physicians have begun by being the physician of the poor.

In William B. Bean
Aphorisms from Latham (p. 25)

Physicians, who have worthily achieved great reputation, become the refuge of the hopeless, and earn for themselves the misfortune of being expected to cure incurable diseases.

In William B. Bean
Aphorisms from Latham (p. 25)

Physicians are in a manner often called upon to be wiser than they possibly can be. Disease or imperfection of a vital organ is a fearfully interesting thing to him who suffers it, and he presses to learn all that is known, and often much more than is known about it.

In William B. Bean
Aphorisms from Latham (p. 26)

The physician pursues the disease through its own channels. He tracks it to its springhead, and takes hold of it there, and puts an end to it.

In William B. Bean
Aphorisms from Latham (p. 70)

The philosophical physician is evermore studying how, upon adequate grounds, he can assign to medical facts this relation. But he knows in how delicate and difficult a task he is engaged. He is obliged to wait upon experience, and to attend to phenomena as they happen to occur. He cannot bring them together at will, and vary and transpose them as he likes, so as to learn their connection. He envies the ease with which the chemist can bring any substance within the sphere and influence of as many others as he pleases; and the accuracy with which he can then ascertain the degrees of affinity it bears severally to each,—an accuracy so precise, that he can express them by numbers.

In William B. Bean
Aphorisms from Latham (p. 90)

The end of all the thought and labour of physicians is to make experiments with men's lives.

In William B. Bean
Aphorisms from Latham (p. 91)

The highest praise which the world has to bestow upon the physician is that he is experienced. There must, therefore, be a good deal worth knowing about this experience, which is deemed his characteristic excellence; as, how he goes to work in search of it, and how he gains it, experimenting after his manner, and with whatever help of science he can muster, or none at all; but still experimenting.

In William B. Bean
Aphorisms from Latham (p. 92)

We physicians had need be a self-confronting and a self-reproving race; for we must be ready, without fear or favor, to call in question our own Experience and to judge it justly; to confirm it, to repeal it, to reverse it, to set up the new against the old, and again to reinstate the old and give it preponderance over the new.

In William B. Bean
Aphorisms from Latham (pp. 93–4)

Longfellow, Henry Wadsworth
You behold in me
Only a traveling Physician;
One of the few who have a mission
To cure incurable diseases,
Or those that are called so.

The Works of Henry Wadsworth Longfellow
Volume V
Christus
The Golden Legend
Part I (p. 144)

Luke 4:23
Physician, heal thyself.

The Bible

Massinger, Philip
1 Doct. What art can do, we promise; physic's hand
As apt is to destroy as to preserve,
If Heaven make not the med'cine: all this while,
Our skill hath combat hell with his disease;
But 'tis so arm'd, and a deep melancholy,
To be such in part with death, we are in fear
The grave must mock our labours.

The Plays of Philip Massinger
Volume I
The Virgin-Martyr
Act IV, Scene I (p. 76)

Pesc. Your physicians are
Mere voice, and no performance; I have found
A man that can do wonders . . . but leaves him
To work this miracle.

The Plays of Philip Massinger
Volume I
The Duke of Milan
Act V, Scene II (p. 340)

Matthew 9:12
They that be whole need not a physician, but they that are sick.

The Bible

Mayo, Charles H.
The true physician will never be satisfied just to pass his therapeutic wares over a counter.

Collected Papers of the Mayo Clinic & Mayo Foundation
Problems in Medical Education
Volume 18, 1926

Mencken, H.L.
The true physician does not preach repentance; he offers absolution.

Prejudices: Third Series
Chapter XIC (p. 269)

Meyer, Adolf
I wonder how soon we shall be far enough along to have the physician ask: How much and what, if anything, is *structural*? how much *Functional, somatic* or *metabolic*? How much *constitutional, psychogenic* and *social*?

New England Journal of Medicine
The "Complaint" As the Center of Genetic–Dynamic and
Nosological Teaching In Psychiatry
August 23, 1928

Montaigne, Michel de
Who ever saw one physician approve of another's prescription, without taking something away, or adding something to it?

Essays
Book the Second
Chapter 37 (p. 371)

Moore, Merrill
If the average man is a harp on whom Nature occasionally plays, the physician is an instrument on whom the emotions are played continuously during his waking hours and that is not too good for any man.

In Mary Lou McDonough
Poet Physicians
Afterthought (p. 198)

More, Hannah
I used to wonder why people should be so fond of the company of their physician, till I recollected that he is the only person with whom one dares to talk continually of oneself, without interruption, contradiction or censure; I suppose that delightful immunity doubles their fees.

Letter to Horace Walpole
July 27, 1789

Osler, Sir William
Permanence of residence, good undoubtedly for the pocket, is not always best for wide mental vision in the physician.

Aequanimitas, with Other Addresses
The Army Surgeon (p. 101)

To wrest from nature the secrets which have perplexed philosophers of all ages, to track to their sources the causes of disease, to correlate the vast stores of knowledge, that they may be quickly available for the prevention and cure of disease—these are our ambitions.

Aequanimitas, with Other Addresses
Chauvinism in Medicine (p. 267)

To prevent disease, to relieve suffering and to heal the sick—this is our work.

> *Aequanimitas, with Other Addresses*
> Chauvinism in Medicine (p. 267)

It may be well for a physician to have pursuits outside his profession, but it is dangerous to let them become too absorbing.

> In Harvey Cushing
> *The Life of Sir William Osler*
> Volume I (p. 67)

To investigate the causes of death, to examine carefully the condition of organs, after such changes have gone on in them as to render existence impossible and to apply such Knowledge to the prevention and treatment of disease, is one of the highest objects of the Physician . . .

> In Harvey Cushing
> *The Life of Sir William Osler*
> Volume I (p. 85)

'Tis no idle challenge which we physicians throw out to the world when we claim that our mission is of the highest and of the noblest kind, not alone in curing disease but in educating the people in the laws of health, and in preventing the spread of plagues and pestilences . . .

> In Harvey Cushing
> *The Life of Sir William Osler*
> Volume I (p. 408)

No class of men needs friction so much as physicians; no class gets less. The daily round of busy practitioners tends to develop an egoism of a most intense kind, to which there is no antidote. The few setbacks are forgotten, the mistakes are often buried, and ten years of successful work tend to make a man touchy, dogmatic, intolerant of correction, and abominably self-centered. To this mental attitude the medical society is the best corrective, and a man misses a good part of his education who does not get knocked about a bit by his colleagues in discussions and criticisms . . .

> In Harvey Cushing
> *The Life of Sir William Osler*
> Volume I (p. 447)

A physician who does not use books and journals, who does not need a library, who does not read one or two of the best weeklies and monthlies, soon sinks to the level of the cross-counter prescriber, and not alone in practice, but in those mercenary feelings and habits which characterize a trade . . .

> In Harvey Cushing
> *The Life of Sir William Osler*
> Volume I (p. 448)

Few men live lives of more devoted self-sacrifice than the family physician but he may become so completely absorbed in work that leisure is unknown . . . There is danger in this treadmill life lest he lose more than health and time and rest—his intellectual independence. More than most men he feels the tragedy of isolation—that inner isolation so well expressed in Matthew Arnold's line—'We mortal millions live *alone*.' Even in populous districts the practice of medicine is a lonely road which winds up-hill all the way and a man may easily go astray and never reach the Delectable Mountains unless he early finds those shepherd guides of which Bunyan tells, *Knowledge, Experience, Watchful* and *Sincere*. The circumstances of life mould him into a masterful, self-confident, self-centered man, whose worst faults often partake of his best qualities.

In Harvey Cushing
The Life of Sir William Osler
Volume I (p. 588)

The physician who shows in his face the slightest alteration, expressive of anxiety or fear, has not his medullary centres under the highest control, and is liable to disaster at any moment. I have spoken of this to you on many occasions, and have urged you to educate your nerve centres so that not the slightest dilator or contractor influence shall pass to the vessels of your face under any professional trial.

Student Life (p. 37)

Ovid
'Tis not always in a physician's power to cure the sick.

Ex Ponto
Book I, iii

Owen, John
Physicians take Gold, but seldom give:
They Physick give, take none; yet healthy live.
A Diet They prescribe; the Sick must for't
Give Gold; Each other Thus supply-support.

Latine Epigrams
Book I, Number 53

Paracelsus
The book of Nature is that which the physician must read; and to do so he must walk over the leaves.

Encyclopædia Britannica
Volume xviii, Ninth edn (p. 234)

Percival, Thomas
The relations in which a physician stands to his patients, to his brethren, and to the public, are complicated, and multifarious; involving much knowledge of human nature, and extensive moral duties.

Medical Ethics (p. viii)

Hospital PHYSICIANS and SURGEONS should minister to the sick, with due impressions of the importance of their office; reflecting that the ease, the health, and the lives of those committed to their charge depend on their skill, attention, and fidelity.

Medical Ethics (p. 9)

Petronius
Medicus nihil aliud est quam animi consolatio.
[A physician is nothing but a consoler of the mind.]

The Satyricon
Section 42

Piozzi, Hester Lynch
A physician can sometimes parry the scythe of death, but has no power over the sand in the hourglass.

Letter to Fanny Burney
November 12, 1781

Plato
. . . no physician, in so far as he is a physician, considers his own good in what he prescribes, but the good of his patient; for the true physician is also a ruler having the human body as a subject, and is not a mere moneymaker.

The Republic
Book I [342] (p. 303)

The most skillful physicians are those who, from their youth upwards, have combined with the knowledge of their art the greatest experience of disease; they had better not be robust in health, and should have had all manner of diseases in their own persons.

The Republic
Book III [408] (p. 337)

. . . so too in the body the good and healthy elements are to be indulged and the elements of disease are not to be indulged, but discouraged. And this is what the physician has to do, and in this the art of medicine consists: for medicine may be regarded generally as the knowledge of the loves and desires of the body, and how to satisfy them or not; and the best physician is he who is able to separate fair love from foul, or to convert one into the other; and he who knows how to eradicate and how to implant love, whichever is required, and can reconcile the most hostile elements in the constitution and make them loving friends, is a skillful practitioner.

Symposium
[186] (p. 156)

Plautus

When sickness comes call for the physician.

Amphitruo
Fragment 12

Pliny the Elder

But for these Physitians, who are the judges themselves to determine of our lives, and who many times are not long about it, but give us a quick dispatch and send us to heaven or hell; what regard is there had, what inquiry and examination is made of their quality and worthiness?

Natural History
XXIX, i

The medical profession is the only one in which anybody professing to be a physician is at once trusted, although nowhere else is an untruth more dangerous.

Natural History
XXIX, viii, 17

Plutarch

. . . a skillful physician, who, in a complicated and chronic disease, as he sees occasion, at one while allows his patient the moderate use of such things as please him, at another while gives him keen pains and drugs to work the cure.

The Lives of the Noble Grecians and Romans
Pericles (p. 129)

Poe, Edgar Allan

Is there—*is* there balm in Gilead?—tell me—tell me, I implore!

The Raven
Stanza 15

Prior, Matthew

I sent for Ratcliffe; was so ill
   That other doctors gave me over:
He felt my pulse—prescrib'd his pill,
   And I was likely to recover.
But when the wit began to wheeze,
   And wine had warm'd the politician

Cur'd yesterday of my disease,
  I died last night of my physician.

In Helen & Lewis Melville's
An Anthology of Humorous Verse
The Remedy Worse than the Disease

Proverb
Where there are three physicians, there are two atheists.

In Oliver Wendell Holmes
Medical Essays
The Medical Profession in Massachusetts (p. 364)

A disobedient patient makes an unfeeling physician.

In Robert Christy
Proverbs, Maxims and Phrases of All Ages (p. 255)

Deceive not thy Physitian, Confessor, nor Lawyer.

In George Herbert
Outlandish Proverbs
#105

God heales, and the Physitian hath the thankes.

In George Herbert
Outlandish Proverbs
#169

Go not for every grief to the Physitian, nor for every quarrel to the Lawyer, nor for every thirst to the pot.

In George Herbert
Outlandish Proverbs
#290

An old Physitian, and a young Lawyer.

In George Herbert
Outlandish Proverbs
#648

There are more Physitians in health than drunkards.

In George Herbert
Outlandish Proverbs
#903

The Physitian owes all to the patient, but the patient owes nothing to him but a little money.

In George Herbert
Outlandish Proverbs
#921

Proverb, Chinese
The physician can cure the sick, but he cannot cure the dead.

> In Robert Christy
> *Proverbs, Maxims and Phrases of All Ages* (p. 259)

Proverb, German
A physician is an angel when employed, but a devil when one must pay him.

> In Robert Christy
> *Proverbs, Maxims and Phrases of All Ages* (p. 256)

When you call a physician call the judge to make your will.

> In Robert Christy
> *Proverbs, Maxims and Phrases of All Ages* (p. 259)

Proverb, Hebrew
Do not dwell in a city whose governor is a physician.

> Source unknown

Proverb, Italian
From your confessor, lawyer and physician,
Hide not your case on no condition.

> In Sir John Harrington
> *Metamorphosis of Ajax*
> The Second Section (p. 154)

A multiplicity of laws and of physicians in a country is equally a sign of its bad condition.

> Source unknown

Dove non va il sole, va il medico
[Where the sunlight enters not, there goes the physician]

> In Robert Means Lawrence
> *Primitive Psycho-Therapy and Quackery*
> The Blue-Glass Mania (p. 95)

Proverb, Japanese
Better go without medicine than call in an unskillful physician.

> Source unknown

Proverb, Latin
Fingunt se medicos quivis idiota, sacerdos, Judæus, momachus, histrio, rasor, anus.
[Every idiot, priest, Jew, monk, actor, barber, and old woman, fancy themselves physicians.]

> Source unknown

Quarles, Francis
Physicians of all men are most happy; what good success soever they
have, the world proclaimeth, and what faults they commit, the earth
coverth.

Hieroglyphikes
iv. Nicocles

Rabelais, François
Happy is the physician, whose coming is desired at the declension of a
disease.

Pantagruel
Book 3, Chapter 41 (p. 209)

Ray, John
Every man is either a fool or a physician after thirty years of age.
A Complete Collection of English Proverbs (p. 30)

Piss clear, and defy the physician.
A Complete Collection of English Proverbs (p. 35)

St. Augustine
. . . one who has tried a bad physician fears to trust himself with a good
one . . .

Confessions
VI [IV]

Scott, Sir Walter
. . . a slight touch of the cynic in manner and habits, gives the physician,
to the common eye, an air of authority which greatly tends to enlarge
his reputation.

The Surgeon's Daughter
Chapter I (p. 23)

. . . the sick chamber of the patient is the kingdom of the physician.
The Talisman
Chapter VII (p. 99)

The praise of the physician . . . is the recovery of the patient.
The Talisman
Chapter VIII (p. 112)

Seegal, David
The young physician today is so generously provided with a kit of
diagnostic and therapeutic tools, his attention might be wisely directed
to the question of "what not to do" as well as "what to do."

Journal of Chronic Diseases
Volume 17, 1964 (p. 299)

Selden, John

Preachers say, do as I say, not as I do. But if the physician had the same disease upon him that I have, and he should bid me to do one thing, and he do quite another, could I believe him?

Table Talk of John Selden
Preaching #13 (p. 145)

Seneca

The physician cannot prescribe by letter . . . he must feel the pulse.

Ad Lucilium Epistulae Morales
Volume I
Epistle xxii, Section 1

Shakespeare, William

Trust not the physician;
His antidotes are poison, and he slays
More than you rob.

Timon of Athens
Act IV, Scene III, L. 434–436

Kill thy physician, and the fee bestow
Upon thy foul disease.

King Lear
Act I, Scene I, L. 164–165

Sheridan, Richard Brinsley

. . . I had rather follow you to your grave, than see you owe your life to any but a regular bred physician.

St. Patrick's Day
Act II
Scene Justice Hoofe (p. 24)

Simmons, Charles

The art of the physician consists, in a great measure, in exciting hope, and other friendly passions and feelings.

Laconic Manual and Brief Remarker (p. 330)

Smollett, Tobias

The character of a physician, therefore, not only presupposes natural sagacity, and acquired erudition, but it also implies every delicacy of sentiment, every tenderness of nature, and every virtue of humanity.

Sir Launcelot Greaves
Chapter XXIV (p. 192)

Stanton, Elizabeth Cady
Besides the obstinacy of the nurse, I had the ignorance of the physicians
to contend with.

Eighty Years and More
Motherhood (pp. 118–9)

Stevenson, Robert Louis
There are men and classes of men that stand above the common herd; the
soldier, the sailor, and the shepherd not unfrequently; the artist rarely;
rarelier still, the clergyman; the physician almost as a rule. He is the
flower (such as it is) of our civilization; and when that stage of man is
done with, and only remembered to be marveled at in history, he will be
thought to have shared as little as any in the defects of the period, and
most notably exhibited the virtue of the race. Generosity he has, such
as is possible to those who practice an art, never to those who drive a
trade; discretion, tested by a hundred secrets; tact, tried in a thousand
embarrassments; and what are more important, Herculean cheerfulness
and courage. So it is that he brings air and cheer into the sick-room, and
often enough, though not so often as he wishes, brings healing.

Underwoods
Preface

Swift, Jonathan
Physicians ought not to give their Judgment of Religion, for the same
Reason that Butchers are not admitted to be Jurors upon Life and Death.

Satires and Personal Writings
Thoughts
On Various Subjects (p. 410)

Taylor, Jeremy
. . . to preserve a man alive in the midst of so many chances, and
hostilities, is as great a miracle as to create him . . .

Holy Living and Holy Dying
Volume II
Chapter I, Section 1, L. 7–9

Thoreau, Henry David
Priests and physicians should never look one another in the face. They
have no common ground, nor is there any to mediate between them.
When the one comes, the other goes. They could not come together
without laughter, or a significant silence, for the one's profession is a
satire on the other's, and either's success would be the other's failure.

A Week on the Concord and Merrimack Rivers
Wednesday (p. 227)

It is wonderful that the physician should ever die, and that the priest should ever live. Why is it that the priest is never called to consult with the Physician? It is because men believe practically that matter is independent of spirit. But what quackery? *It is commonly an attempt to cure the disease of a man* by addressing his body alone. There is a need of a physician who shall minister to both soul and body at once, that is to man. Now he falls between two stools.

A Week on the Concord and Merrimack Rivers
Wednesday (p. 227)

Unknown
A member of the faculty in a London medical college was appointed an honorary physician to the king. He proudly wrote a notice on the blackboard in his classroom:

"Professor Jennings informs his students that he has been appointed honorary physician to His Majesty, King George."

When he returned to the class-room in the afternoon he found written below his notice this line:

"God save the King."

In Edward J. Clode
Jokes for All Occasions
Doctors (p. 73)

Nature, time and patience are the three great physicians.

Source unknown

A new physician must have a new churchyard.

In Robert Burton
The Anatomy of Melancholy
Volume II
Part 2, Section 4 (p. 36)

In illness the physician is a father; in convalescence a friend; when health is restored, he is a guardian.

Source unknown

No good physician ever takes physic.

Source unknown

That city is in a bad case whose physician has the gout.

Source unknown

The disobedience of the patient makes the physician seem cruel.

Source unknown

The physician is often more to be feared than the disease.

Source unknown

The four best physicians, Dr. Sobriety, Dr. Jocosity, Dr. Quiet, and Dr. Gold.

Source unknown

The physician cannot drink the medicine for the patient.

Source unknown

When the physician can advise the best the patient is dead.

Source unknown

Villanova, Arnald of
. . . the physician must be learned in diagnosing, careful and accurate in prescribing, circumspect and cautious in answering questions, ambiguous in making prognosis, just in making promises; and he should not promise health because in doing so he would assume a divine function and insult God.

In Henry E. Sigerist (Translator)
Quarterly Bulletin of Northwestern University Medical School
Bedside Manners in the Middle Ages: The Treatise De Cautelis Medicorum
Attributed to Arnald of Villanova
Volume 20, 1946

Virchow, Rudolf
Only those who regard healing as the ultimate goal of their efforts can, therefore, be designated as physicians.

Disease, Life, and Man
Standpoints in Scientific Medicine (p. 26)

Voltaire
Let nature be your first physician.
It is she who made all.

Philosophical Dictionary
Volume VII
Medicine (p. 169)

But nothing is more estimable than a physician who, having studied nature from his youth, knows the properties of the human body, the diseases which assail it, the remedies which will benefit it, exercises his art with caution, and pays equal attention to the rich and the poor.

Philosophical Dictionary
Volume VIII
Physicians (pp. 199–200)

The Devil should not try his tricks on a clever physician. Those familiar
with nature are dangerous for the wonder-workers. I advise the Devil
always to apply to the faculty of theology—not to the medical faculty.

Annals of Medical History
Volume 1, 1917
Quoted in Pearch Bailey
Voltaire's Relation to Medicine (p. 58)

von Ebner-Eschenbach, Marie
Physicians are hated either on principle or for financial reasons.

Aphorisms (p. 50)

Webster, John
Physicians are like kings—,
they brook no contradiction.

The Duchess of Malfi
Act V, Scene II, L. 69-70

Wordsworth, William
Physician art thou? one, all eyes,
Philosopher! a fingering slave,
One that would peep and botanize
Upon his mother's grave

The Complete Poetical Works of William Wordsworth
A Poet's Epitaph

Young, Arthur
. . . there is a great difference between a good physician and a bad one;
yet very little between a good one and none at all.

Travels in France
September 9, 1787 (p. 66)

PHYSIOLOGIST

Bayliss, William Maddock
To use an illustration, inadequate as it is, that of a petrol motor, the problem of the physiologist is analogous to that of the investigation of the amount of fuel consumed in relation to the amount of work done, when the engine is working under various conditions. The greater number of the chemical and physical properties of the materials used in the construction of the engine are of no importance, such as the valency of the iron or the smell of the lubricating oil, while others are fundamental, such as the heat of combustion of the fuel and the insulation of the ignition circuit. Even the exact chemical nature of the fuel is of a subsidiary importance, so long as it is sufficiently volatile, and capable of giving an explosive mixture with oxygen. Moreover, the precise form of many parts, such as the heads of bolts, is immaterial, just as many structural details of living organisms or the precise chemical composition of connective tissue have, at all events at present, an insignificant physiological interest.

Principles of General Physiology
Preface (pp. xv–xvi)

Collingwood, R.G.
A physiologist . . . can certainly offer a definition of life; but this will only be *an interim report on the progress of physiology to date*. For him, as for the beginner, it is the nature of physiology that is relatively certain; it is the nature of life that is relatively vague.

The New Leviathan
Part 1, Chapter 1, aphorism I.47

Mayo, Charles H.
Disease at times creates experiments that physiology completely fails to duplicate, and the wise physiologist can obtain clues to the resolution of many problems by studying the sick.

Annals de Circulation
La funcion del higado en relacion con la cirugia
Volume 2, April 1930

PHYSIOLOGY

Huxley, Thomas Henry
There is no side of the human mind which physiological study leaves uncultivated. Connected by innumerable ties with abstract science, Physiology is yet in the most intimate relation with humanity, and by teaching us that law and order, and a definite scheme of development, regulate even the strangest and wildest manifestations of individual life, she prepares the student to look for a goal even amidst the erratic wanderings of mankind, and to believe that history offers something more than an entertaining chaos—a journal of a toilsome, tragi-comic march nowhither.

Collected Essays
Volume III
Science and Education (p. 59)

A thorough study of Human Physiology is, in itself, an education broader and more comprehensive than much that passes under that name. There is no side of the intellect which it does not call into play, no region of human knowledge into which either its roots, or its branches, do not extend; like the Atlantic between the Old and the New Worlds, its waves wash the shores of the two worlds of matter and of mind; its tributary streams flow from both; through its waters, as yet unfurrowed by the keel of any Columbus, lies the road, if such there be, from the one to the other; far away from that North-west Passage of mere speculation, in which so many brave souls have been hopelessly frozen up.

Collected Essays
Volume III
Science and Education (p. 220)

PILL

Crichton-Browne, Sir James
If you want fame and fortune, invent a pill.

The Doctor's After Thoughts (p. 14)

Fuller, Thomas
If the pills were pleasant, they would not want gilding.

Gnomologia
Number 2711

Garth, Samuel
Some fell by *Laudanum*, and some by *Steel*,
And Death in Ambush lay in ev'ry Pill.

The Dispensary
Canto IV, L. 62–63

Herrick, Robert
When his potion and his pill
His, or none, or little skill
Meet for nothing, but to kill;
    Sweet Spirit comfort me!

The Complete Poems of Robert Herrick
Volume III
His Litanie, to the Holy Spirit (p. 132)

Jerrold, Douglas
A pill that the present moment is daily bread to thousands.

The Catspaw
Act I, Scene I

Molière
My lord Jupiter knows how to gild the pill.

Amphitryon
Act III, Scene X, L. 24

Ray, John
Apothecaries would not give pills in sugar unless they were bitter.

A Complete Collection of English Proverbs (p. 2)

Shakespeare, William
When I was sick, you gave me bitter pills.

The Two Gentlemen of Verona
Act II, Scene IV, L. 149

Stumpf, LaNore
How is it that a little pill
Without a pair of eyes to see
Can travel down, and round and round
And figure out what's wrong with me?

American Journal of Nursing
Needed: Remote Control (p. 902)
April 1969

Unknown
Bitter pills are gilded.

Source unknown

Bitter pills may have wholesome effects.

Source unknown

Pills must be bolted not chewed.

Source unknown

PIMPLE

Siegel, Eli
A pimple has atoms to it; and mucus has electrons.

Damned Welcome
Aesthetic Realism Maxims
Part II, #234 (p. 121)

PMS

Bates, Rhonda

My doctor said "I've got good news and I got bad news. The good news is you don't have Premenstrual Syndrome. The bad news is—you're a bitch!"

In Roz Warren
Glibquips (p. 122)

Hankla, Susan

God grant me the serenity to change the things about me and others
    I cannot stand
And to stand the things about me and others I cannot change
And the insight to know the difference
Between a PMS day and a normal day
So no one gets hurt.

In Roz Warren
Glibquips (p. 122)

PNEUMONIA

Glasow, Arnold

Spring fever in some areas is double pneumonia.

Quote, The Weekly Digest
June 18, 1967 (p. 497)

Unknown

There was a young man from Laconia
Whose mother-in-law had pneumonia
   He hoped for the worst,
   And after March first,
They buried her 'neath a begonia.

In Louis Untermeyer
Lots of Limericks (p. 37)

PRAYER

Clowes, William

O Almighty God and most merciful father of our Lord Jesus Christ, we most humbly acknowledge before thy glorious Majesty, that by our manifold sins and continual transgressions of thy laws and commandments, we most justly procure thy heavy displeasure against us, and provoke thy Majesty not only to plague us with grievous sickness and diseases of our body, but also to bring the most dreadful horror and terror of eternal damnation, and the torments of everlasting death upon body and soul for ever. But thou are the Father of mercy and the God of all comfort and wouldest not the death of a sinner. Thou are the Heavenly Physician, that hast not only provided but also proffered to miserable man the wholesome medicines of health and deliverance for body and soul. Have mercy upon us therefore, loving Father; pardon and forgive us all our sins and wickedness, and grant us daily more and more to apply unto our souls the most comfortable medicine of the holy word, that we may thereby increase in a true and lively faith and a sound knowledge of thy holy will. Make it profitable unto us, good Lord, to spy out all our spiritual sickness and diseases, and to find the true remedies for the same, that we may fly from the occasions that may draw us to sin, and recover strength more and more, against all our several sins and corruptions. And for as much as thou hast also graciously provided outward remedies for the diseases of our bodies and appointed Physicians and Surgeons, the ministers of the same, we beseech thee, make us diligent in searching, careful in using, and faithful in practising and applying of those remedies that thou has taught us. Bless our labours, we beseech thee, that thy power giving force to these medicines, they may be effectual to the removing the griefs of thy people. And grant that both we and they that shall receive help by us may hereby earnestly be stirred up to praise and magnify thy holy name . . .

Selected Writings of William Clowes
A Prayer (pp. 168–9)

Corinthians, Second Epistle 1:3–4
Blessed be God, the Father of mercies, and the God of all comfort, who comforteth us in all our tribulations; that we may be able to comfort those who are in trouble, with the comfort wherewith we ourselves also are comforted of God.

The Bible

Lederman, Leon
Dear Lord, forgive me the sin of arrogance, and Lord, by arrogance I mean the following . . .

In John D. Barrow
The Artful Universe (p. 31)

Lewis, Sinclair
God give me the unclouded eyes and freedom from haste. God give me quiet and relentless anger against all pretense and all pretentious work and all work left slack and unfinished. God give me a restlessness whereby I may neither sleep nor accept praise till my observed results equal my calculated results or in pious glee I discover and assault my error. God give me the strength not to trust to God.

Arrowsmith
Chapter XXVI, Section II (p. 292)

Percival, Thomas
And I devoutly pray that the blessing of God may attend all your pursuits; rendering them at once subservient to your own felicity, and the good of your fellow-creatures.

Medical Ethics (p. ix)

Stewart, George
When I, having finished with things below,
Lie out'neath the sod alone,
Raise no cold monument to me
Of brass or bronze or stone;
But plant me beneath a big oak tree,
With its roots firmly fixed in the sod,
And its branches pointing everywhere
To the throne of the living God.
Plant me out in the open on some fair hill
And not in a burying ground;
'Twould be hard for me I know to keep still
With other folk lying around:
And if you can manage a glimpse of the Bay,
Where the white sails shake in the sun,
I know I'll be far more willing to stay—

I always did love a gun,
But I s'pose that shooting will be forbid
Unless you've got the pull,
Why? on account of the brimstone, Kid!
That stuff's inflammable.

In Mary Lou McDonough
Poet Physician
A Tired Doctor's Prayer (p. 128)

Make a mixture of onions, lard and cornmeal . . .
and then place it on the persons chest.
Unknown – (See p. 303)

PRESCRIPTION

Chekhov, Anton
"What did your uncle die of?"

"Instead of fifteen Butkin drops, as the doctor prescribed, he took sixteen."

Note-Book of Anton Chekhov (p. 37)

Helmuth, William Tod
Term pain "neuralgia," of if the man be stout,
Cry out, "Dear Sir, you have *rheumatic* gout."
Tap on the chest—some *awful* sounds they hear,
Then satisfied, declare, "The case is clear,"
Draw forth a paper, seize the magic quill,
And write in mystic signs, "*Cathartic pill*."

Scratches of a Surgeon
Medical Pomposity (p. 11)

Latham, Peter Mere
To bring many important remedies together, and unite them by a lucky combination, and compress them within a small compass, and so place them within the common reach, all this gives a facility of prescribing which is hurtful to the advance of medical experience. The facility of prescribing is a temptation to prescribe; and, under this temptation, there is a lavish expenditure continually going on of important remedies in the mass, of which the prescribers have made no sufficient experiment in detail.

In William B. Bean
Aphorisms from Latham (p. 60)

Thoreau, Henry David
There are sure to be two prescriptions diametrically opposite. Stuff a cold and starve a cold are but two ways.

A Week on the Concord and the Merrimack River
Wednesday

Twain, Mark

It would be a good thing for the world at large, however unprofessional it might be, if medical men were required by law to write out in full the ingredients named in their prescriptions. Let them adhere to the Latin, or Fejee, if they choose, but discard abbreviations, and form their letters as if they had been to school one day in their lives, so as to avoid the possibility of mistakes on that account.

San Francisco Morning Call
Damages Awarded
10/1/1864

Unknown

There's only one thing harder to read than the handwriting on the wall, and that's a doctor's prescription.

In Evan Esar
20,000 Quips and Quotes

Where ignorance is bliss, 'tis folly for a doctor to tell a patient what he has written on his prescription.

In Evan Esar
20,000 Quips and Quotes

Wynter, Dr.

Tell me from whom, fat-headed Scot,
  Thou didst thy system learn;
From Hippocrates thou hadst it not,
  Nor Celsus, nor Pitcairn.
Suppose that we own that milk is good,
  And say the same of grass;
The one for babes is only food,
  The other for an ass.
Doctor! our new prescription try
  (A friend's advice forgive);
Eat grass, reduce thyself, and die;—
  Thy patients then may live.

In William Davenport Adams
English Epigrams
On Doctor Cheyne, the Vegetarian
cclxxvi

PROGNOSIS

Hoffmann, Friedrich
There is nothing better than a skillful prognosis to commend a physician
or to win him a reputation.

Fundamenta Medicianae
Semiotics
Chapter 4, 1 (p. 96)

In making a prognosis the physician should proceed cautiously, for he
can easily lose his reputation by a rash prognosis.

Fundamenta Medicianae
Semiotics
Chapter 4, 2 (p. 96)

PROTEIN

Byrne, Josefa Heifetz
methionylglutaminylarginyltyrosylglutamylserylleucylphenylalanylalanylglutamin
ylleucyllysylglutamylarginyllysylglutamylglycylalanylphenylalanylvalylprolyl
phenylalanylvalylthreonylleucylglycylaspartylprolylglycylisoleucylglutamylglu
taminylserylleucyllysylisoleucylaspartylthreonylleucylisoleucylglutamylalanyl
glycylalanylaspartylalanylleucylglutamylleucylglycylisoleucylprolylphenylala
nylserylaspartylprolylleucylalanylaspartylglycylprolylthreonylisoleucylgluta
minylasparaginylalanylthreonylleucylarginylalanylphenylalanylalanylalanylgly
cylvalylthreonylprolylalanylglutaminylcysteinylphenylalanylglutamylmethionyl
leucylalanylleucylisoleucylarginylglutaminyllysylhistidylprolylthreonylisoleu
cylprolylisoleucylglycylleucylleucylmethionyltyrosylalanylasparaginylleucylva
lylphenylalanylasparaginyllysylglycylisoleucylaspartylglutamylphenylalanyltyro
sylalanylglutaminylcysteinylglutamyllysylvalylglycylvalylaspartylserylvalylleu
cylvalylalanylaspartylvalylprolylvalylglutaminylglutamylserylalanylprolylphe
nylalanylarginylglutaminylalanylalanylleucylarginylhistidylasparaginylvalylala
nylprolylisoleucylphenylalanylisoleucylcysteinylprolylprolylaspartylalanylas
partylaspartylaspartylleucylleucylarginylglutaminylisoleucylalanylseryltyrosyl
glycylarginylglycyltyrosylthreonyltyrosylleucylleucylserylarginylalanylglycyl
valylthreonylglycylalanylglutamylasparaginylarginylalanylalanylleucylprolylleu
cylasparaginylhistidylleucylvalylalanyllysylleucyllysylglutamyltyrosylasparagi
nylalanylalanylprolylprolylleucylglutaminylglycylphenylalanylglycylisoleucylse
rylalanylprolylaspartylglutaminylvalyllysylalanylalanylisoleucylaspartylalanyl
glycylalanylalanylglycylalanylisoleucylserylglycylserylalanylisoleucylvalylly
sylisoleucylisoleucylglutamylglutaminylhistidylasparaginylisoleucylglutamylpro
lylglutamyllysylmethionylleucylalanylalanylleucyllysylvalylphenylalanylvalyl
glutaminylprolylmethionyllysylalanylalanylthreonylarginylserine,
n.: The chemical name for *tryptophan synthetase A protein*, a 1,913-letter
enzyme with 267 amino acids.

Mrs. Bryne's Dictionary of Unusual, Obscure, and Preposterous Words

QUACK

Bernstein, Al

You can usually tell a quack doctor by his bill.

Quote, The Weekly Digest
July 28, 1968 (p. 77)

Bierce, Ambrose

QUACK, *n*. A murderer without a license.

The Enlarged Devil's Dictionary

Bishop, Samuel

When quacks, as quacks may by good luck, to be sure,
Blunder out at haphazard a desperate cure,
In the prints of the day, with due pomp and parade,
Case, patient, and doctor are amply display'd.
And this is quite just—and no mortal can blame it;
If they save a man's life, they've a right to proclaim it,
But there's reason to think they might save more lives still,
Did they publish a list of the numbers they kill!

In William Davenport Adams
English Epigrams
Audi Alteram Partem
cclxxxv

Clowes, William

Yea, nowadays, it is too apparent to see how tinkers, tooth-drawers, peddlers, ostlers, carters, porters, horse-gelders and horse-leeches, idiots, apple-squires, broom-men, bawds, witches, conjurers, soothsayers and sow-gelders, rogues, ratcatchers, runagates and proctors of Spittlehouses, with such other like rotten and stinking weeds do in town and country, without order, honesty or skill, daily abuse both Physic and Surgery, having no more perseverance, reason or knowledge in this art than has a goose, but only a certain blind practice, without wisdom or judgment, and most commonly use one remedy for all diseases and one way of curing to all persons, both old and young, men, and women and children, which is as possible to perform or to be true as for a shoemaker with one last to make a shoe fit for every man's foot, and this is one principal cause that so many perish.

Selected Writings of William Clowes
Of Blind Buzzards and Cracking Cumbatters (pp. 77–8)

Colton, Charles Caleb

It is better to have recourse to a quack, if he can cure the disorder, although he cannot explain it, than to a physician, if he can explain our disease, but cannot cure it.

Lacon
1:170

Crabbe, George

A potent quack, long versed in human ills,
Who first insults the victim whom he kills; . . .

The Poetical Works of George Crabbe
The Village, L. 282–283 (p. 15)

Graves, Richard

A doctor, who, for want of skill,
Did sometimes cure—and sometimes kill;
Contriv'd at length, by many a puff,
And many a bottle fill'd with stuff,
To raise his fortune, and his pride;
And in a coach, forsooth! must ride.
His family coat long since worn out,
What arms to take, was all the doubt.
A friend, consulted on the case,
Thus answer'd with a sly grimace:
"Take some device in your own way,
Neither too solemn nor too gay;
Three ducks, suppose; white, grey, or black;
And let your motto be, Quack! quack!"

In William Davenport Adams
English Epigrams
A Doctor's Motto
cclxxxi

Hood, Thomas

Not one of these self-constituted saints,
Quacks—not physicians—in the cure of souls.

The Poetical Works of Thomas Hood
Volume 1
Ode to Rae Wilson Esq., L. 14–15

Jenner, Edward
I've dispatch'd, my dear madam, this scrap of a letter,
To say that Miss ____ is very much better.
A Regular Doctor no longer she lacks,
And therefore I've sent her a couple of Quacks.

In William Davenport Adams
English Epigrams
Sent to a Patient, with the Present of a Couple of Ducks
cclxxiii

Jonson, Ben
Pergrine: There are quack salvers,
Fellows that live by venting oils and drugs.

Volpone
Act II, Scene II, L. 56–57

Massinger, Philip
Out, you impostors!
Quacksalving, cheating mountebanks! your skill
Is to make sound men sick, and sick men kill.

The Plays of Philip Massinger
Volume I
The Virgin-Martyr
Act IV, Scene I (p. 78)

Unknown
Paulus, the famous quack, renown'd afar,
For killing more than pestilence or war,
Of late, in orders, is a curate made,
And buries people—not to change his trade.

In William Davenport Adams
English Epigrams
On a Certain Quack
ccxxvi

Wycherley, William
A quack is as fit for a pimp as a midwife for a bawd: they are still but
in their way, both helpers of nature.

The Country Wife
Act 1 (p. 5)

RADIOLOGIST

Unknown

The radiologists' favorite journal is the *Wall Street Journal*.

Source unknown

RAPPORT

Unknown

"How do you establish rapport with a patient?" was the question posed by a young nursing student as she pored over the notes she had taken in a class she describes as "new concepts in nursing." The person she was asking was her mother who has been in nursing for twenty-five-years and is now a busy recovery room head nurse. Mama's reply was so immediate, it was practically a reflex: "If you have to ask, you've already lost it!"

Quote, The Weekly Digest
July 17, 1966 (p. 9)

RECOVERY

Byron, Lord George Gordon
Despair of all recovery spoils longevity,
And makes men's miseries of alarming brevity.

Don Juan
Canto II, Stanza LXIV

Massinger, Philip
O my Doctor,
I never shall recover.

The Bondman
Act I, Scene I

REMEDY

Defoe, Daniel
As frightened Patients, when they want a cure,
Bid any Price, and any Pain endure:
But when the Doctor's Remedies appear
The Cure's too Easy, and the Price too Dear.

Selected Poetry and Prose of Daniel Defoe
The True-Born Englishman
Part II
Britannia (p. 70)

Dunne, Finley Peter
Ivery gineration iv doctors has had their favrite remedies.

Mr. Dooley Says
Drugs (p. 97)

Florio, John
Unto al is remedie, except unto death.

Firste Fruites
Proverbs, Chapter 19

Hippocrates
For extreme diseases, extreme methods of cure . . .

Aphorisms
Section I, 6

Holmes, Oliver Wendell
Many things are uncertain in this world, and among them the effect of a large proportion of the remedies prescribed by physicians.

Guardian Angel
XI (p. 130)

Latham, Peter Mere

You cannot be sure of the success of your remedy, while you are still uncertain of the nature of the disease.

In William B. Bean
Aphorisms from Latham (p. 61)

To scatter above twenty remedies, and let hit which may, is like pigeon-shooting in companies. The bird falls; but whose gun was it that brought it down? Nobody is reputed the better marksman after a hundred volleys.

In William B. Bean
Aphorisms from Latham (p. 61)

I conceive it hardly possible for a physician to employ his time worse than in quest of new specifics. His common sense would be about equal to that of the man who should trust his hopes of growing rich to the chance of finding a bag of money. But a specific medicine is an excellent thing, and so is a bag of money; and, being found, it is worth the study of a life to turn them both to all the good purposes of which they are capable.

In William B. Bean
Aphorisms from Latham (p. 61)

Syrus, Publilius

There are some remedies worse than the disease.

Sententiae
Number 301

Twain, Mark

[A reply to letters recommending remedies]: Dear Sir (or Madam):
— I try every remedy sent to me. I am now on No. 67. Yours is 2,653. I am looking forward to its beneficial results.

In Clara Clemens
My Father Mark Twain
One April Evening (p. 287)

REMEDY: BLEEDING

Unknown

A gob of spider webs and soot placed over a cut is supposed to stop the bleeding.

REMEDY: CHICKEN POX

Unknown

Sitting under a chicken roost may bring about a quick cure for someone with chicken pox.

REMEDY: COLDS

Fleming, Sir Alexander
A good gulp of hot whiskey at bedtime—it's not very scientific, but it helps.

Suggestion on treatment of the common cold
March 22, 1954

Gordon, Richard
On the first sign of a cold, go to bed with a bottle of whiskey and a hat. Place hat on left-hand bedpost. Take a drink of whiskey and move hat to right-hand post. Take another drink and shift it back again. Continue until you drink the whiskey but fail to move the hat. By then the cold is probably cured.

Atlantic Monthly
The Common Cold
January 1955

Simmons, Charles
Starve a fever, and cram a cold.

Laconic Manual and Brief Remarker (p. 87)

Unknown
Drink a cup of peppermint tea and munch a clove of garlic to help a cold.

Drink a cup of regular tea to which has been added a teaspoon of butter and cinnamon and a teaspoon each of honey and whiskey.

Gargle with tobasco sauce, six to ten drops, in a cup of warm water.

Make a mixture of onions (chopped and fried in grease), lard and cornmeal. Place this mixture on a red flannel cloth, warm in the oven and then place it on the person's chest.

You won't get a cold if you smear your entire body with bear grease.

If you kiss a mule, you will quickly cure a bad cold.

REMEDY: FAINTING

Unknown
A faint bottle that is made of camphor and whiskey is good to have on hand to wave under the nose of someone who faints.

REMEDY: FEVER

Unknown
For fever, take one teaspoon of salt mixed in water. After that put a spoonful of salt inside each stocking as soon as you feel a chill coming on.

Chew some willow bark to reduce a fever.

Place a damp lemon peel under each armpit for a low grade fever.

REMEDY: GERMS

Unknown
Peeled onions hung in a room where sickness is will help kill the germs.

REMEDY: HEADACHE

Unknown
Soak a towel in vinegar and wrap it around your head for a migraine headache.

Place a fresh lemon peel across your forehead to relieve the pain of a headache.

To relieve the pain of a headache, drink a half cup of tea with a shot of whiskey and honey in it.

REMEDY: MEASLES

Unknown
Drinking buttermilk liberally laced with black pepper helps get measles bumps out on a patient and then keep them out.

REMEDY: MENSES

Unknown
Drinking black pepper tea can start a late period.

REMEDY: NAUSEA

Unknown
Place a copper penny on your navel to treat nausea.

To help deter nausea try sipping a mixture of Jell-O and water.

REMEDY: NOSEBLEED

Unknown
Ice placed on the nape of the neck will stop a nosebleed.

REMEDY: SPRAINS

Unknown
Vinegar-soaked brown paper placed on a sprain will ease pain and help with the healing.

REMEDY: WARTS

Unknown
Picking up a rock and spitting under it can get rid of warts.

To get rid of warts, rub it with raw meat and then feed the meat to a dog.

RESEARCH

Bates, Marston
Research is the process of going up alleys to see if they are blind.

In Jefferson Hane Weaver
The World of Physics
Volume II (p. 63)

Green, Celia
The way to do research is to attack the facts at the point of greatest astonishment.

The Decline and Fall of Science
Aphorisms (p. 1)

Lasker, Albert D.
"Research," he said, "is something that tells you that a jackass has two ears."

In John Gunther
Taken at the Flood: The Story of Albert D. Lasker (p. 96)

Mizner, Wilson
If you steal from one author, it's plagiarism; if you steal from many, it's research.

In Alva Johnston
The Legendary Mizners
Chapter 4, The Sport (p. 66)

Smith, Theobald
Research is fundamentally a state of mind involving continual re-examination of the doctrines and axioms upon which current thought and action are based. It is therefore critical of existing practices. Research is not necessarily confined to the laboratory and complex instruments and apparatus which require laboratory housing. This is controlled research which endeavors to pick out of the web of nature's activities some single strand and trace it toward its origin and its terminus

and determine its relation to other strands. The older type of research involving observation and study of the entire fabric of disease largely with the help of the unaided senses, such as was the practice of doctors a century ago, has had its day, but backed by experience and a keen observant mind it even now occasionally triumphs over the narrow controlled research of the laboratory. It is the kind used by Darwin and other early biologists in establishing on a broad, comparative basis, the evolution of plant and animal life.

American Journal of the Medical Sciences
The Influence of Research in Bringing into Closer Relationship the
Practice of Medicine and Public Health Activities (p. 19)
December 1929

It is not infrequently taken for granted that the results of laboratory research are final and its operations infallible. This, however, is far from true. The laws, theories, and inferences of experimental research are as subject to rectification as are inferences based on the other human activities. They are approximations getting nearer and nearer the actuality with time.

American Journal of the Medical Sciences
The Influence of Research in Bringing into Closer Relationship the
Practice of Medicine and Public Health Activities (p. 24)
December 1929

Szent-Györgyi, Albert
Research means going out into the unknown with the hope of finding something new to bring home. If you know what you are going to do, or even to find there, then it is not research at all, then it is only a kind of honourable occupation.

In Jefferson Hane Weaver
The World of Physics
Volume II (p. 63)

von Braun, Wernher
Basic research is when I'm doing what I don't know I'm doing.

In Jefferson Hane Weaver
The World of Physics
Volume II (p. 63)

von Ebner-Eschenbach, Marie
When curiosity turns to serious matters, it's called research.

Aphorisms (p. 26)

REST

Crichton-Browne, Sir James

Rest in bed is in itself the cure of many maladies . . . Many patients take to it readily, as a retreat from the turmoil of the hour, but others protest against it and say: "Why should I not dress and lie on the sofa with a rug over me, which is just the same thing?" But it is not the same thing, for there is in rest in bed an *abandon* and sense of tranquillity which the best upholstered sofa cannot afford, and it gives overstrained organs like the heart time to rally and recuperate.

From the Doctor's Notebook
Rest in Bed (p. 21)

RESUSCITATE

II Kings 4:32–34

And when E-lisha was come into the house, behold, the child was dead, and laid upon his bed.

He went in therefore, and shut the door upon them twain, and prayed unto the Lord.

And he went up, and lay upon the child, and put his mouth upon his mouth, and his eyes upon his eyes, and his hands upon his hands: and he stretched himself upon the child; and the flesh of the child waxed warm.

The Bible

RHEUMATISM

Esar, Evan
Rheumatism. The only thing that keeps more people on the straight and narrow path than conscience.

Esar's Comic Dictionary

Haskins, Henry S.
Much more is known about the stars than about rheumatism.

In Herbert V. Prochnow and Herbert V. Prochnow, Jr
A Treasury of Humorous Quotations

von Ebner-Eschenbach, Marie
One believes in rheumatism and true love only when afflicted by them.

Aphorisms (p. 49)

SALVE

Fuller, Thomas

Different sores must have different salves.

Gnomologia
Number 1283

Ray, John

There's a *salve* for every sore.

A Complete Collection of English Proverbs (p. 156)

Unknown

A salve there is for every sore.

School-House of Women
L. 401

SCIENCE

Allbutt, Sir Thomas Clifford
In science, law is not a rule imposed from without, but an expression of an intrinsic process.

Quoted by F.H. Garrison
Bulletin of the New York Academy of Medicine
Volume 4, 1928 (p. 1000)

Arnauld, Antoine
We are accustomed to use reason as an instrument for acquiring the sciences, but we ought to use the sciences as an instrument for perfecting the reason.

The Art of Thinking: The Port-Royal Logic
First Discourse (p. 7)

Bernard, Claude
. . . in science, the best precept is to alter and exchange our ideas as fast as science moves ahead.

An Introduction to the Study of Experimental Medicine
Part I, Chapter II, Section IV (p. 41)

Science increases our power in proportion as it lowers our pride.

Bulletin of the New York Academy of Science
Volume 64, 1928 (p. 997)

A contemporary poet has characterized this sense of the personality of art and of the impersonality of science in these words,—"Art is myself; science is ourselves."

An Introduction to the Study of Experimental Medicine
Part I, Chapter II, Section IV (p. 43)

True science suppresses nothing, but goes on searching and is undisturbed in looking straight at things that it does not yet understand.

> *An Introduction to the Study of Experimental Medicine*
> Part III, Chapter IV, Section IV (p. 223)

Science goes forward only through new ideas and through creative or original power of thought.

> *An Introduction to the Study of Experimental Medicine*
> Part III, Chapter IV, Section IV (p. 226)

Boulding, Kenneth E.

Science might almost be defined as the process of substituting unimportant questions which can be answered for important questions which cannot.

> *The Image*
> Chapter 11 (p.164)

The important thing in science is not so much to obtain new facts as to discover new ways of thinking about them.

> In A. Koestler and J.R. Smithies
> *Beyond Reductionism*

Browne, Sir Thomas

No one should approach the temple of science with the soul of a money-changer . . .

> In Harvey Cushing
> *The Life of Sir William Osler*
> Volume I (p. 129)

da Costa, J. Chalmers

Every science is a chain composed of many links. A link represents a truth and often bears the name of the torchbearer who found the truth. There are, of course, nameless links because the names of the discoverers are dim in the twilight of fable or are hidden in the blackness of prehistoric chaos.

> *The Trials and Triumphs of the Surgeon*
> Francis X. Dercum (p. 172)

de Madariaga, Salvadore

. . . medical science can only be based on the direct observation of the patient as a living whole, made by a concrete and all-round doctor possessing, of course, the required theoretical and practical knowledge but, above all, inborn gifts and trained qualifications as an observer.

> *Essays with a Purpose*
> On Medicine (pp. 175–6)

Mayo, William J.

The wit of science not only expresses but actually reveals the science and art of medicine.

Journal of the American Medical Association
In the Time of Henry Jacob Bigelow
Volume 77, August 20, 1921 (p. 599)

Naunyn, Bernhard

Our patients obeyed us gladly. Our zeal led them to respect and trust us. It never occurred to them to inquire whether this zeal was in the interest of treatment or in the interest of science.

Deutsches Archiv für klinische Medizin
140 Bd., 1 u. 2 H. (p. 27)

Osler, Sir William

Let no man be so foolish as to think that he has exhausted any subject for his generation. Virchow was not happy when he saw the young men pour into the old bottles of cellular pathology the new wine of bacteriology. Lister could never understand how aseptic surgery arose out of his work. Ehrlich would not recognize his epoch-making views on immunity when this generation has finished with them. I believe it was Hegel who said that progress is a series of negations—a denial today of what was accepted yesterday, the contradiction by each generation of some part at least of the philosophy of the last; but all is not lost, the germ plasma remains, a nucleus of truth to be fertilized by men, often ignorant even of the body from which it has come. Knowledge evolves, but in such a way that its possessors are never in sure possession.

Evolution of Modern Medicine
Chapter VI (p. 219)

To the physician particularly a scientific discipline is an incalculable gift, which leavens his whole life, giving exactness of habits of thought and tempering the mind with that judicious faculty of distrust which can alone, amid the uncertainties of practice, make him wise unto salvation.

Aequanimitas
The Leaven of Science (p. 92)

Penfield, Wilder

The trouble is not in science but in the uses men make of it. Doctor and layman alike must learn wisdom in their employment of science, whether this applies to atom bombs or blood transfusion.

The Second Career
A Doctor's Philosophy

Ramón y Cajal, Santiago
Science, like life, grows ever, renewing itself continually without running against the wall of decrepitude in its creative impetus.

Recollections of My Life
Chapter XXVIII (p. 604)

Stern, B.J.
. . . sciences seemingly as remote as astronomy have modified the attitude of physician and patient toward disease by transforming man's conception of his place in the universe. Astrology had to go before scientific medicine could ever be conceived . . . Mathematics has aided, not merely in description but in prognosis and prevention of disease. The development of biology gave important clues to anatomy and physiology, and botany's advance gave new scope to *materia medica*.

Chemistry has not only provided new drugs but also made possible new concepts of disease such as dietary deficiency and glandular dysfunction. It has revolutionized medical knowledge of the functioning of the human organism . . . Now psychology and sociology, relative latecomers, stand on the threshold, seeking recognition as within the aegis of essential medical knowledge.

Society and Medical Progress
Chapter X (p. 217)

SICK

Dunlap, William

He seems a little under the weather, somehow; and yet he's not sick.

The Memoirs of a Water Drinker
Volume I
Chapter VIII (p. 80)

Dunne, Finley Peter

. . . whin a man's sick, he's sick an' nawthin' will cure him or annything will.

Mr. Dooley Says
Drugs (p. 97)

Emerson, Ralph Waldo

It is dainty to be sick, if you have leisure and convenience.

Journals
Volume V (p. 162)

Fuller, Thomas

He who was never sick dies the first fit.

Proverb
In Thomas Fuller
Gnomologia

Halsted, Anna Roosevelt

There are so many indignities to being sick and helpless . . .

In Joseph P. Lash
Eleanor: The Years Alone
To The End, Courage
Letter to David Gray, November 1, 1962 (p. 327)

Harvey, Gabriel

Now Sicke as a dog . . .

Works
Volume I
Foure Letters and Certaine Sonnets (p. 161)

Herold, Don
Some of the pleasantest, dreamiest, most profitable days of my life have been those I have spent sick in bed. Give me sickness and a day, and I will make the pomp of emperors look ridiculous.

The Happy Hypochondriac (p. 33)

Isaiah 1:5
. . . the whole head is sick, and the whole heart faint.

The Bible

Jerome, Jerome K.
It is a curious fact, but nobody ever is seasick—on land. At sea, you come across plenty of people very bad indeed, whole boat-loads of them; but I never met a man yet, on land, who had ever known at all what it was to be seasick.

Three Men in a Boat
Chapter 1 (p. 10)

Johnson, Samuel
It is so very difficult for a sick man not to be a scoundrel.

Miscellanies
Volume I (p. 267)

What can a sick man say, but that he is sick?

Boswell's Life of Johnson
iv, 362

Luttrell, Henry
Come, come, for trifles never stick:
  Most servants have a failing;
Yours, it is true, are sometimes sick,
  But mine are always ale-ing.

In William Davenport Adams
English Epigrams
On Ailing and Ale-ing
dclxxiii

Shakespeare, William
What, is Brutus sick,
And will he steal out of his wholesome bed,
To dare the vile contagion of the night?

Julius Caesar
Act II, Scene I, L. 263–265

Sterne, Laurence
I am sick as a horse . . .

Tristram Shandy
Volume VII, Chapter II (p. 4)

Swift, Jonathan
Poor Miss, she's sick as a Cushion . . .

The Prose Works of Jonathan Swift
Volume the Fourth
Polite Conversation
Dialogue I (p. 153)

Unknown
The worst thing about being sick is having a disease you can't afford.

In Evan Esar
20,000 Quips and Quotes

In literature a man is ill, but in real life he is sick.

In Evan Esar
20,000 Quips and Quotes

Wolfe, Thomas
Most of the time we think we're sick it's all in the mind.

Look Homeward, Angel
Part I, Chapter 1 (p. 10)

SICK BED

Lamb, Charles

If there be a regal solitude, it is a sick-bed. How the patient lords it there; what caprices he acts without control! how kinglike he sways his pillow-tumbling, and tossing, and shifting, and lowering, and thumping, and flatting, and moulding it, to the ever-varying requisitions of his throbbing temples.

Essays of Elia
The Convalescent (pp. 329–30)

SICKNESS

Barnes, Djuna

No man needs curing of his individual sickness; his universal malady is what he should look to.

Nightwood
La Somnambule (p. 41)

Burton, Robert

Sickness is the mother of modesty, putteth us in minde of our mortality; and, when wee are in the full careere of worldly pompe and jollity, she pulleth us by the eare, and maketh us knowe ourselves.

The Anatomy of Melancholy
Volume II
Part 2, Section 3 (p. 11)

Chrysotome

Princes, Masters, Parents, Magistrates, Judges, Friends, Eniemies, faire or foule meanes cannot containe us; but a little sicknesse will correct and amend us.

In Robert Burton
The Anatomy of Melancholy
Volume II
Part 2, Section 3 (p. 1351)

Donne, John

And can there be worse sickness, than to know
That we are neuer well, nor can be so?

An Anatomy of the World
The First Anniversary, L. 93–94

Emerson, Ralph Waldo

For sickness is a cannibal which eats up all the life and youth it can lay hold of, and absorbs its own sons and daughters.

Complete Works
Volume 6
The Conduct of Life (p. 262)

Fuller, Thomas

Sickness is felt, but health not at all.

Gnomologia
Number 4160

Harris, Joel Chandler

We er sorter po'ly, Sis Tempy, I'm 'blige ter you. You know w'at de jay-bird say ter der squinch-owls!, 'I'm sickly but sassy'.

Nights with Uncle Remus
Chapter 50 (p. 298)

Hood, Thomas

I'm sick of gruel, and the dietics,
I'm sick of pills, and sicker of emetics,
I'm sick of pulse, tardiness or quickness,
I'm sick of blood, its thinness or its thickness,—
In short, within a word, I'm sick of sickness!

The Poetical Works of Thomas Hood
Volume 2
Fragment (p. 424)

Jonson, Ben

Take heed, sickness, what you do,
I shall fear you'll surfeit too.
Live not we, as all they stalls,
Spittles, pest-house, Hospitals,
Scarce will take our present store?

In Robert Bell
The Poems of Robert Greene, Christopher Marlowe, and Ben Jonson
The Forest
VIII. To Sickness

Lamb, Charles

How sickness enlarges the dimensions of a man's self to himself! he is his own exclusive object. Supreme selfishness is inculcated upon him as his only duty.

Essays of Elia
Last Essays of Elia
The Convalescent (p. 330)

Milton, John
. . . all maladies
Of ghastly Spasm, or racking torture, qualms
Of heart-sick Agonie, all feverous kinds,
Convulsions, Epilepsies, fierce Catarrhs,
Intestine Stone and Ulcer, Colic pangs,
Dropsies and Asthmas, and Joint-racking Rheums.

Paradise Lost
Book XI, L. 480–485

O'Connor, Flannery
I have never been anywhere but sick. In a sense sickness is a place, more instructive than a long trip to Europe, and it's always a place where there's no company, where nobody can follow. Sickness before death is a very appropriate thing and I think those who don't have it miss one of God's mercies.

The Habit of Being

Proverb
The chamber of sickness is the chapel of devotion.

Source unknown

Sickness tells us what we are.

Source unknown

Roy, Gabrielle
The Christian Scientists held that it was not God Who wanted sickness, but man who puts himself in the way of suffering. If this were the case, though, wouldn't we all die in perfect health?

The Cashier
Chapter 3 (pp. 36–7)

Shakespeare, William
This sickness doth infect
The very life-blood of our enterprise.

The First Part of King Henry the Fourth
Act IV, Scene I, L. 28–29

Sickness is catching.

A Midsummer-Night's Dream
Act I, Scene I, L. 186

My long sickness
Of health and living now begins to mend,
And nothing brings me all things.

Timon of Athens
Act V, Scene I, L. 189–191

Unknown
Study sickness while you are well.

Source unknown

To have and to hold from this day forward, for better for worse, for richer
for poorer, in sickness and in health . . .

Book of Common Prayer: Solemnization of Matrimony

Weingarten, Violet
Sickness, like sex, demands a private room, or at the very least, a discrete
curtain around the ward bed.

Intimations of Mortality (p. 3)

SINUS

Lamport, Felicia
I hereby confess
That of all I possess
I'd most gladly be minus
The sinus.

Scrap Irony
Lines on an Aching Brow

SKELETON

Esar, Evan

[Skeleton] A lot of bones with the person scraped off.

Esar's Comic Dictionary

Thompson, D'Arcy

The 'skeleton', as we see it in a Museum, is a poor and even a misleading picture of mechanical efficiency. From the engineer's point of view, it is a diagram showing all the compression-lines, but by no means all of the tension-lines of the construction; . . . it falls all to pieces unless we clamp it together, as best we can, in a more or less clumsy and immobilized way. In preparing or 'macerating' a skeleton, the naturalist nowadays carries on the process till nothing is left but the whitened bones. But the old anatomists . . . were wont to macerate by easy stages; and in many of their most instructive preparations, the ligaments were intentionally left in connection with the bones, and as part of the 'skeleton'.

On the Nature and Action of Certain Ligaments (1884)

Unknown

A skeleton is a man with his inside out and his outside off.

In Alexander Abingdon
Bigger & Better Boners (p. 68)

One use of the skeleton is if you once sit down without it you couldn't stand up.

In Alexander Abingdon
Bigger & Better Boners (p. 68)

SKIN

Eliot, T.S.

Webster was much possessed by death
And saw the skull beneath the skin . . .

The Complete Poems and Plays 1909–1950
Whispers of Immortality

Sherman, Alan

Skin is what you feel at home in
And without it, furthermore
Both your liver and abdomen
Would keep falling on the floor
(And you'd be dressed in your intestine).

Source unknown
You Gotta Have Skin

SNEEZE

Carroll, Lewis
Speak roughly to your little boy,
  And beat him when he sneezes:
He only does it to annoy,
  Because he know it teases.

The Complete Works of Lewis Carroll
Alice's Adventures in Wonderland
Chapter 6 (p. 68)

Dimmick, Edgar L.
To sneeze, or not to sneeze; that is the question.
Whether it is nobler in the mind to suffer
The stings and lachrymation of outrageous hay fever
Or to take shots against a sea of troubles
And by much needling end them? To cry, to sneeze
No more, and by a sneeze to end
The headache and the thousand devilish symptoms
The flesh is heir to; 'tis a consummation devoutly to be wished.
To cry, to sneeze,
To sneeze, perchance to stream; ay, there's the rub
For in that sneeze allergic when it comes what streams
May run from noses and from eyes
Must give us pause.

Obstetrics and Gynecology
Volume 20, Number 1, 1962 (p. 148)

Montaigne, Michel de
Will you ask me, whence comes the custom of blessing those who sneeze?
We break wind three several ways; that which sallies from below is too
filthy; that which breaks out from the mouth carries with it some reproach

of having eaten too much; the third eruption is sneezing, which because it proceeds from the head, and is without offence, we give it this civil reception . . .

Essays
Book the Third
Chapter 6 (p. 435)

There is nothing better than a skillful prognosis . . .
Friedrich Hoffmann – (See p. 293)

SORE THROAT

Friedman, Shelby
Sore throat—Strep tease.

Quote, The Weekly Digest
May 7, 1967 (p. 377)

Kraus, Jack
SORE THROAT: Hoarse and buggy.

Quote, The Weekly Digest
August 14, 1966 (p. 17)

Nye, Bill
If your throat gets inflamed, a doctor asks you to run your tongue out into society about a yard and a half, and he pries your mouth open with one of Rogers Brothers' spoon handles. Then he is able to examine your throat as he would a page of the *Congressional Record*, and to treat it with some local application.

Remarks
Spinal Meningitis (p. 123)

SORES

Shakespeare, William
For to strange sores strangely they strain the cure.

Much Ado About Nothing
Act IV, Scene I, L. 254

You rub the sore,
When you should bring the plaster.

The Tempest
Act II, Scene I, L. 138–139

SPECIALIST

Herodotus
Medicine is practiced among them on a plan of separation; each physician treats a single disorder, and no more: thus the country swarms with medical practitioners, some undertaking to cure diseases of the eye, others of the head, others again of the teeth, others of the intestines, and some those which are not local.

The History of Herodotus
The Second Book, 84

Mayo, Charles H.
The definition of a specialist as one who 'knows more and more about less and less' is good and true. Its truth makes essential that the specialist, to do efficient work, must have some association with others who, taken altogether, represent the whole of which the specialty is only a part.

Modern Hospital
Volume 51, September 1938

Ogilvie, Sir Heneage
Medicine as a whole is too vast for the grasp of any one individual. Specialism is inevitable, and having accepted it we must examine its limitations. The essential and inescapable one is that a specialist is expert for one purpose only. A specialist alone can be supremely efficient, and the earlier he devotes himself to a particular branch of study, the smaller that branch, and the more single-minded the devotion with which he studies it, the more efficient will he be in that branch and that alone. But disease is no specialist. Patients do not consult us because certain organs are affected, but because they feel ill. They come with symptoms, and the earlier and therefore the more curable their malady is, the more vague will those symptoms be, the more difficult the elucidation of their cause, the greater the need, in the first place, of a general investigation by one whose daily practice covers the whole of disease.

The Lancet
A Surgeon's Life (p. 3)
Volume 255, July 3, 1948

Unknown

A specialist is a doctor who has discovered which of his talents will bring in the most money.

Source unknown

Choose your specialist and you choose your disease.

The Westminster Gazzette
May 18, 1906

You can usually tell a quack doctor by his bill.
Al Bernstein – (See p. 295)

SPINE

Esar, Evan

[Spinal Column] A collection of bones running up and down your back that keeps your legs from reaching your neck.

Esar's Comic Dictionary

Unknown

The spine is a long line of bones that holds you together. Your head sets on one end and you set on the other.

In Alexander Abingdon
Bigger & Better Boners (p. 68)

STATISTICS

Paulos, John Allen
It's easier and more natural to react emotionally than it is to deal dispassionately with statistics or, for that matter, with fractions, percentages, and decimals. The media (actually, all of us) frequently solve this problem by leaving numbers out of stories and hiding behind such evasive words as "many" or "uncommon", which are almost completely devoid of meaning.

A Mathematician Reads the Newspaper
Cellular Phones Tied to Brain Cancer (pp. 80–1)

Health statistics may be bad for our mental health. Inundated by too many of them, we tend to ignore them completely, to react to them emotionally, to accept them blithely, to disbelieve them closed-mindedly, or simply to misinterpret their significance.

A Mathematician Reads the Newspaper
Ranking Health Risks: Experts and Laymen Differ (p. 133)

Peacock, E.E.
One day when I was a junior medical student, a very important Boston surgeon visited the school and delivered a great treatise on a large number of patients who had undergone successful operations for vascular reconstructions. At the end of the lecture, a young student at the back of the room timidly asked, "Do you have any controls?" Well, the great surgeon drew himself up to his full height, hit the desk, and said, "Do you mean did I not operate on half of the patients?" The hall grew very quiet then. The voice at the back of the room very hesitantly replied, "Yes, that's what I had in mind." Then the visitor's fist really came down as he thundered, "Of course not. That would have doomed half of them to their death." It was absolutely silent then, and one could scarcely hear the small voice ask, "Which half?"

Medical World News
September 1, 1972 (p. 45)

STETHOSCOPE

Dickinson, Richard W.
In order for the stethoscope to function, two things have to happen. *There has to be, by God, a sick man at one end of it and a doctor at the other!* The doctor has to be within thirty inches of his patient.

Transactions of the Association of American Physicians
Volume 75, Number 1, 1962

Holmes, Oliver Wendell
There was a young man in Boston town,
  He bought himself a stethoscope nice and new,
All mounted and finished and polished down,
  With an ivory cap and a stopper too.

The Complete Poetical Works of Oliver Wendell Holmes
The Stethoscope Song

STOMACH

Hunter, William

Some physiologists will have it that the stomach is a mill, others that is a fermenting vat, others again that is a stewpan; but, in my view of the matter, it is neither a mill, a fermenting vat nor a stewpan but a stomach, gentlemen, a stomach.

In Herbert J. Muller
Science and Criticism
Chapter V (p. 106)

STOOLS

Chatton, Milton J.
Our thanks to frank Doctor Duane
Who takes the time to explain
Just how he had noted
That his stools often floated
Before they were flushed down the drain.

He must have thought first, "Mama mia!
Do I suffer from steatorrhea?
But it cannot be that—
There is no trace of fat."
Which led to another idea.

Well aware of the gas he unloosed
The doctor quite shrewdly deduced,
(Almost clairvoyant)
His feces were buoyant
Because of the methane produced.

The New England Journal of Medicine
Methanosis (p. 362)
Volume 287, Number 2, 1972

Hazlitt, William Carew
A Joyner on a time took a pill, and it so wrought with him, that he had fourty stooles in a minute on houre.

Shakespeare Jest Books
Volume III
Conceit, Clichés, Flashes and Whimzies
Number 183

Teller, Joseph D.
While safe's the stool that comes a sinker,
The floater's apt to be a stinker.

So it's not fat, but rather, flatus
Imparts the elevated status.

The New England Journal of Medicine
Floaters and Sinkers (p. 52)
Volume 287, Number 1, 1972

BLESS YOU!

. . .whence comes the custom of blessing those who sneeze?. . .
Michel de Montaigne – (See p. 327)

SURGEON

Aylett, Robert
For *Mercy* doth like skilfull Surgeon deal,
That hath for ev'ry sore a remedy:
If gentle drawing plaisters cannot heal
The wound, because it festreth inwardly,
He sharper corrasives must then apply,
And as he oft cuts off some member dead,
Or rotten, lest the rest should putrifie,
So *Mercy* wicked Members off doth shred,
Lest they should noysome prove to body and the head.

Peace with Her Foure Gardners
The Brides Ornaments
Meditation III, L. 307–315

Caldwell, George W.
Who is the man in sterile white
Delving deep at the point of light,
With nurses, trained, at left and right?
   The Surgeon.

In Mary Lou McDonough
Poet Physician
The Surgeon (p. 136)

Celsus, Aulus Cornelius
A surgeon ought to be young, or at any rate, not very old; his hand should be firm and steady, and never shake; he should be able to use his left hand with as much dexterity as his right; his eye-sight should be acute and clear; his mind intrepid, and so far subject to pity as to make him desirous of the recovery of his patient, but not so far as to suffer himself to be moved by his cries; he should neither hurry the

339

operation more than the case requires, nor cut less than is necessary, but do everything just as if the other's screams made no impression upon him.

In Samuel Evans Massengill
A Sketch of Medicine and Pharmacy (p. 30)

Crichton-Browne, Sir James
Every great surgeon, it used to be said, shakes, swears or sweats when he operates.

The Doctor Remembers
Bret Harte (p. 170)

"Oh, sir, we made a terrible mistake in the case of that man yesterday! We amputated the wrong leg!"

"Ah well," the surgeon replied, complacently, "it's of no consequence, for I have just been looking at the other leg, and it's going to get better."

The Doctor's After Thoughts (p. 15)

Croll, Oswald
. . . it is necessary that every Surgeon should be a Physitian, and every Physitian a Chyrugion, that there may be a sound Bridegroom for a sound Bride . . .

Philosophy Reformed and Improved in Four Profound Tractates (p. 151)

Cvikota, Raymond J.
Surgeon: Fee lancer.

Quote, The Weekly Digest
June 9, 1968 (p. 457)

da Costa, J. Chalmers
A surgeon is like a postage stamp. He is useless when stuck on himself.

The Trials and Triumphs of the Surgeon
The Trials and Triumphs of the Surgeon (p. 17)

A vain surgeon is like a milking stool; of no use except when sat upon.

The Trials and Triumphs of the Surgeon
The Trials and Triumphs of the Surgeon (p. 17)

Davies, Robertson
A man mentioned casually to me this afternoon that his brother was in a hospital, having his appendix removed. This operation is now undertaken without qualm; surgeons regard it as a pastime, something to keep the hands busy, like knitting or eating salted nuts.

The Table Talk of Samuel Marchbanks (p. 176)

de Chauliac, Guy
The surgeon should be learned, skilled, ingenious, and of good morals. Be bold in things that are sure, cautious in dangers; avoid evil cures and practices; be gracious to the sick, obliging to his colleagues, wise in his predictions. Be chaste, sober, pitiful, and merciful; not covetous nor extortionate of money, but let the recompense be moderate, according to the work, the means of the sick, the character of the issue or event, and its dignity.

In Samuel Evans Massengill
A Sketch of Medicine and Pharmacy (p. 262)

de Mondeville, Henri
. . . A Surgeon ought to be fairly bold. He ought not to quarrel before the laity, and although he should operate wisely and prudently, he should never undertake any dangerous operation unless he is sure that it is the only way to avoid a greater danger. His limbs, and especially his hands, should be wellshaped with long, delicate and supple fingers which must not be tremulous.

In John Arderne
Treatise of Fistula in Ano (p. xx)

Dickens, Charles
"What! don't you know what a Sawbones is, Sir?" inquired Mr. Wheeler. "I thought everybody know'd as a Sawbones was a Surgeon."

Pickwick Papers
Volume II
Chapter 30 (p. 413)

Dickinson, Emily
Surgeons must be very careful
When they take the knife!
Underneath their fine incisions
Stirs the Culprit,—Life!

Poems (1890–1896)
XVI

Dimmick, Edgar L.
The public views his status regas,
In the profession he's "The Eagle."
With super-supple fingers slim
(Pus, blood, and guts don't bother him),
Up to his elbows, filled with glee,
With snick and slice sadistically,
Into a jar, up on a shelf

He puts a fragment of yourself.
For him no diagnostic doubt—
He'll operate, and so find out.

Journal of the American Medical Association
The In-Side of Two
Volume 199, Number 6, 1967 (p. 274)

Dunlop, William
There is hardly on the face of the earth a less enviable situation than that of an Army Surgeon after a battle . . .

Recollection of the War of 1812
Chapter III (p. 54)

Eliot, T.S.
The wounded surgeon plies the steel
That questions the distempered part;
Beneath the bleeding hands we feel
The sharp compassion of the healer's art
Resolving the enigma of the fever chart.

Four Quartets
East Coker
Part 4 (p. 15)

Friedman, Shelby
There's the surgeon whose towels are marked HEMORRHAGE and HERNIA.

Quote, The Weekly Digest
February 18, 1968 (p. 137)

Gilbertus, Anglicus
Why in God's name is there such a great difference between a physician and a surgeon?

Quoted in
Time
Surgery (p. 44)
May 3, 1963

Hazlitt, William Carew
An ignorant drunken Surgeon that kil'd all men that came under his hands, boasted himself a better man than the Parson; for, said he, your Cure maintains but yourself, but my Cures maintaine all the Sextons in the Towne.

Shakespeare Jest Books
Volume III
Conceit, Clichés, Flashes and Whimzies
Number 163

Helmuth, William Tod
. . . *doctors* are the *Devil's progeny*,
While surgeons come *directly down from God!*

Scratches of a Surgeon
Surgery vs. Medicine (p. 66)

Kafka, Franz
That is what people are like in my district. Always expecting the
impossible from the doctor. They have lost their ancient beliefs; the
parson sits at home and unravels his vestments, one for another; but the
doctor is supposed to be omnipotent with his merciful surgeon's hands.
Well, as it pleases them; I have not thrust my services on them . . .

The Complete Stories
A Country Doctor (p. 224)

Massinger, Philip
Well.: Thou wert my surgeon; you must tell no tales;
    Those days are done. I will pay you in private.

A New Way to Pay Old Debts
Act IV, Scene II (p. 123)

McCoy, Dr. Leonard
I'm a surgeon, not a psychiatrist.

Star Trek
The City On The Edge Of Forever

Ogilvie, Sir Heneage
A surgeon conducting a difficult case is like the skipper of an ocean-
going racing yacht. He knows the port he must make, but he cannot
foresee the course of the journey.

The Lancet
A Surgeon's Life (p. 1)
Volume 255, July 3, 1948

Paretsky, Sara
Heart surgeons do not have the world's smallest egos: when you ask
them to name the world's three leading practitioners, they never can
remember the names of the other two.

In Marilyn Wallace (Editor)
Sisters in Crime
Volume I
The Case of the Pietro Andromache
II (p. 116)

Proverb, English
The best surgeon is he that hath been hacked himself.

Source unknown

Proverb, Spanish
There is no better surgeon than one with many scars.

Source unknown

Shadwell, Thomas
Oh this Surgeon! this damn'd Surgeon, will this Villanous Quack never come to me? Oh this Plaster on my Neck! It gnaws more than *Aqua-Fortis*: this abominable Rascle has mistaken sure, and given me the same Caustick he appli'd to my Shins, when they were open'd last.

The Complete Works of Thomas Shadwell
The Humorists
The First Act

Twain, Mark
It is a gratification to me to know that I am ignorant of art, and ignorant also of surgery. Because people who understand art find nothing in pictures but blemishes, and surgeons and anatomists see no beautiful women in all their lives, but only a ghastly stack of bones with Latin names to them, and a network of nerves and muscles and tissues.

Mark Twain's Travels with Mr. Brown
Academy of Design (p. 238)

Unknown
An unfortunate lady named Piles
Had the ugliest bottom for miles;
   But her surgeon took pity
   And made it so pretty:
All dimples, and poutings, and smiles.

The Little Limerick Book (p. 32)

Call not a surgeon before you are wounded.

Source unknown

He mistakes the knife of the surgeon for the blade of the assassin.

Source unknown

The best surgeon is he of the soul.

Source unknown

Tender surgeons make foul wounds.

Source unknown

[Surgeon] A butcher with a medical degree.

Esar's Comic Dictionary

[Surgeon] A doctor whose work is cut and dried.

Esar's Comic Dictionary

Surgeons do it.
Internists talk about it.
Radiologists just look at the pictures.

Source unknown

A good surgeon must have an eagle's eye, a lion's heart, and a lady's hand.

Display of Dutie
See also Bulwer, *What will He do with It?*, Book XII, II
and John Ray, *A Complete Collection of English Proverbs* (p. 31)

A chance to cut is a chance to cure.

Source unknown

There are only three rules to life for a surgeon: Eat when you can, sleep when you can, and don't screw with the pancreas.

Source unknown

Don't look for things that you don't want to find.

Source unknown

Never argue with your surgeon: he has inside information.

In Evan Esar
20,000 Quips and Quotes

When a man talks more than a woman about his operations, he is probably a surgeon.

In Evan Esar
20,000 Quips and Quotes

The only bad thing about being on call every other night is that you miss half of the educational opportunities.

Source unknown

SURGERY

Clendening, Logan
Surgery does the ideal thing—it separates the patient from his disease.
Modern Methods of Treatment
Part I, Chapter I (p. 17)

Dennis, F.S.
There is no science that calls for greater fearlessness, courage, and nerve than that of surgery, none that demands more of self-reliance, principle, independence and the determination in the man. These were the characteristics which were chiefly conspicuous in the early settlers of this country. And it is these old-time Puritan qualities, which descending to them in succeeding generations, have passed into surgeons of America, giving them boldness in their art, and enabling them to win that success in surgery, which now commands the admiration of the civilized world.
The History and Development of Surgery during the Past Century (p. 84)

Hubbard, Elbert
SURGERY: An adjunct, more or less valuable to the diagnostician.
The Roycroft Dictionary (p. 143)

Johnson, Ernest
[Back fusions are] like killing a fly on a windowpane with a sledgehammer. The fly is dead, but you've also broken the glass.
Time
That Aching Back (p. 34)
July 14, 1980

Kirklin, John
Surgery . . . is always second best. If you can do something else, it's better. Surgery is limited. It is operating on someone who has no place else to go.

Quoted in
Time
Surgery (p. 60)
May 3, 1963

Mayo, William J.
Surgery is more a matter of mental grasp than it is of handicraftsmanship.

Surgery, Gynecology and Obstetrics
Master Surgeons of America; Frederic S. Dennis
Volume 67, October 1938

O'Malley, Austin
Surgery: by far the worst snob among the handicrafts.

In Herbert V. Prochnow and Herbert V. Prochnow, Jr
A Treasury of Humorous Quotations

Ogilvie, Sir Heneage
Surgery thus attracts the man whose interest in medicine is humanitarian rather than scientific, who loves his fellow men, who wishes to help them and to see that his help is effective. It appeals to the craftsman who enjoys the use of his hands, to the artist whose mind works on visual images, to the romantic who enjoys the drama of life, particularly when it affords him the opportunity to play a decisive role, to the extrovert.

The Lancet
A Surgeon's Life (p. 1)
Volume 255, July 3, 1948

Ovid
The knife must cut the cancer out, infection
Averted while it can be . . .

Metamorphoses
Book I, L. 190–191

Selzer, Richard
Surgery is the red flower that blooms among the leaves and thorns that are the rest of medicine.

Letters to a Young Doctor

One enters the body in surgery, as in love, as though one were an exile returning at least to his hearth, daring uncharted darkness in order to reach home.

Mortal Lessons
The Surgeon as Priest (p. 25)

SURGICAL

Bierce, Ambrose

Before undergoing a surgical operation arrange your temporal affairs. You may live.

The Cynic's Word Book

Mayo, Charles H.

Carry out the two fundamental surgical requirements: see what you are doing and leave a dry field.

Collected Papers of the Mayo Clinic & Mayo Foundation
Volume 27

SYMPTOM

Hare, Hobart Amory
A clear understanding by the physician of the value of the symptoms of disease which he sees and of those described by the patient is of vital importance for the purpose of diagnosis and treatment, and one of the advantages of older physicians over their younger brethren is the ability which they have gained through long training to grasp the essential details of a case almost at their first glance at the patient.

Practical Diagnosis
Introduction (p. 17)

Latham, Peter Mere
It is by symptoms, and by symptoms only, that we can learn the existence, and seat, and nature, of diseases in the living body, or can direct and methodize their treatment.

In William B. Bean
Aphorisms from Latham (p. 59)

TEACHING

Colton, Charles Caleb
Examinations are formidable, even to the best prepared, for the greatest fool may ask more than the wisest man can answer.

Lacon (p. 170)

Ehrensvärd, Gösta Carl Henrik
. . . consciousness will always be one dimension above comprehensibility.

Man on Another World (p. 151)

Eldridge, Paul
What difference does it make how often we lower and raise the bucket into the well if the bucket has no bottom?

Maxims for a Modern Man
484

Fabing, Howard
Marr, Ray
Modern education is a mess. We do not learn for keeps. We do not correlate nor do we pyramid our facts. We do not teach the *how* and *why*, but rather the *what*—which is always changing.

Fischerisms (p. 29)

Hutchison, Sir Robert
Those of us who have the duty of training the rising generation of doctors must not inseminate the virgin minds of the young with the tares of our own fads. It is for this reason that it is easily possible for teaching to be too "up-to-date". It is always well, before handing the cup of knowledge to the young, to wait until the froth has settled.

British Medical Journal
Fashions and Fads in Medicine
1925

Unknown

The instructor in the Medical College exhibited a diagram.

"The subject here limps," he explained, "because one leg is shorter than the other." He addressed one of the students:

"Now, Mr. Snead, what would you do in such a case?"

Young Snead pondered earnestly and replied with conviction:

"I fancy, sir, that I should limp, too."

In Edward J. Clode
Jokes for All Occasions
Doctors (pp. 73–4)

Nil carborundum illegitimi.
[Don't let the bastards grind you down.]

Source unknown

Fallacia plurium interrogationum.
[The fallacy of many questions.]

Source unknown

The Twenty-Third Qualm
The professor is my quizmaster, I shall not flunk.
He maketh me to enter the examination room;
He leadeth me to an alternative seat;
He restoreth my fears.
Yea, though I know not the answers to those questions,
the class average comforts me.
I prepare my answers before me in the sight of my proctors.
I anoint my exam papers with figures.
My time runneth out.
Surely grades and examinations will follow all the days of my life,
And I will dwell in this class forever.

Source unknown

It is not needful in the present day to discourage thinkers, they are not too numerous.

Westminster Review
Exclusion of Opinion (p. 49)
Volume 29, 1838

Wells, H.G.
No man can be a good teacher when his subject becomes inexplicable.

Experiment in Autobiography
Chapter the Fifth (p. 176)

TEETH

Alison, Richard
Those cherries fairly do enclose
Of orient pearl a double row,
Which, when her lovely laughter shows,
They look like rosebuds fill'd with snow.

An Howre's Recreation in Musike

Baxter, Richard
An aching tooth is better out than in.
To lose a rotten member is a gain.

Hypocrisy

Berry, James H.
Brush them and floss them and take them to the dentist. Care for them
and they will stay with you. Ignore them, and they'll go away.

Time
Special Advertising Section (p. 21)
February 11, 1985

Christie, Agatha
Beastly things, teeth . . . Give us trouble from the cradle to the grave.

At Bertram's Hotel
Chapter X (p. 93)

de la Salle, St. Jean Baptiste
It is necessary to clean the teeth frequently, more especially after meals,
but not on any account with a pin, or the point of a penknife, and it
must never be done at table.

The Rules of Christian Manners and Civility
I

Editor of the Louisville Journal
Probably the reason why women's teeth decay sooner than men's is not the perpetual friction of their tongues upon the pearl, but rather the intense sweetness of their lips.

In George Denison Prentice
Prenticeana (p. 35)

Esar, Evan
False teeth may help many a man to keep a stiff upper lip.

20,000 Quips and Quotes

It hurts just as much to have a tooth extracted as it does to have one pulled.

20,000 Quips and Quotes

[Tooth] A painful subject—if not when coming, when going out.

Esar's Comic Dictionary

Exodus 21:24
Leviticus 24:20
Deuteronomy 19:21
Matthew 5:38
. . . tooth for tooth . . .

The Bible

Fabing, Howard
Marr, Ray
I find that most men would rather have their bellies opened for five hundred dollars than have a tooth pulled for five.

Fischerisms

Franklin, Benjamin
Hot things, sharp things, sweet things, old things, all rot the teeth.

Poor Richard's Almanac
1734

Hazlitt, William Carew
One said a tooth drawer was a kind of unconscionable trade, because his trade was nothing else but to take away those things whereby every man gets his living.

Shakespeare Jest Books
Volume III
Conceit, Clichés, Flashes and Whimzies
Number 84

Herrick, Robert
Some ask'd how pearls did grow, and where,
Then spoke I to my girle,
To part her lips, and showed them there
The quarelets of pearl.

Shakespeare Jest Books
Conceit, Clichés, Flashes and Whimzies
Number 84

Hood, Thomas
The best of friends fall out, and so
His teeth had done some years ago.

The Poetical Works of Thomas Hood
Volume 1
A True Story
Stanza 2

Job 19:20
I escaped with the skin of my teeth.

The Bible

Lamb, Charles
The fine lady, or fine gentleman, who show me their teeth, show me bones.

The Complete Works and Letters of Charles Lamb
The Praise of Chimney-Sweeps (p. 99)

Martial
Thais has black, Laecania snowy teeth. What is the reason? One has those she purchased, the other her own.

Epigrams
Book V, Epigram XLIII

Mayo, Charles H.
A crowned tooth is not a 'crown of glory' and may cover a multitude of germs.

Minnesota Medicine
Problems of Infection
Volume 1, 1918

O'Donoghue, Michael
Tough teeth make tough soldiers.

Edited by staff
National Lampoon Tenth Anniversary Anthology
Frontline Dentists (p. 111)

Perelman, S.J.
I'll dispose of my teeth as I see fit, and after they've gone, I'll get along. I started off living on gruel, and by God, I can always go back to it again.

Crazy Like a Fox
Nothing but the Tooth (p. 72)

Proverb, Italian
Chi ha denti, non ha pane; e chi ha pane, non ha denti.
[He that has teeth, has not bread; he that has bread, has not teeth.]

Source unknown

Proverb, Spanish
God gives almonds to those who have no teeth.

Source unknown

My teeth are closer to me than my relatives.

Source unknown

Shakespeare, William
Bid them wash their faces,
And keep their teeth clean.

Coriolanus
Act II, Scene III, L. 65–66

Last scene of all,
That ends this strange eventful history,
Is second childishness and mere oblivion,
Sans teeth, sans eyes, sans taste, sans everything.

As You Like It
Act II, Scene VII, L. 163–166

Skelton, John
In the spyght of his tethe . . .

The Complete English Poems
Why Come Ye Nat to Courte, L. 942

Song of Solomon 4:2
Thy teeth are like a flock of sheep that are even shorn, which came up from the washing.

The Bible

Swift, Jonathan
. . . sweet Things are bad for the Teeth.

The Prose Works of Jonathan Swift
Volume the Fourth
Polite Conversation
Dialogue II (p. 181)

Twain, Mark
Adam and Eve had many advantages, but the principal one was that they escaped teething.

The Tragedy of Pudd'nhead Wilson
Chapter IV

When teeth became touched with decay or were otherwise ailing, the doctor knew of but one thing to do—he fetched his tongs and dragged them out. If the jaw remained, it was not his fault.

Mark Twain's Autobiography

Unknown
Removing the teeth will cure something, including the foolish belief that removing the teeth will cure everything.

Source unknown

A permanent set of teeth consists of 8 canines, 8 cuspids, 2 molars, and 8 cuspidors.

In Alexander Abingdon
Bigger & Better Boners (p. 70)

Wheeler, Hugh
To lose a lover or even a husband or two during the course of one's life can be vexing. But to lose one's teeth is a catastrophe.

A Little Night Music
Act II, Scene I (p. 113)

THEORY

Latham, Peter Mere

It may be doubted whether the theories of clever men have not done more harm to the practice of medicine than all the mere blunders of the ignorant put together.

In William B. Bean
Aphorisms from Latham (p. 100)

TONGUE DEPRESSOR

Armour, Richard
The tongue depressor does the trick
 On old as well as young.
The gadget's really rather slick
 For holding down the tongue.
Although it's somewhat short and flat,
 It's like a magic wand
For making tongues lie down so that
 The view is clear beyond.

The Medical Muse
The Tongue Depressor

TOOTHACHE

Burns, Robert

My curse upon your venom'd stang,
That shoots my tortur'd gooms alang,
An' thro' my lug gies monie a twang
   Wi' gnawing vengeance,
Tearing my nerves wi' bitter pang,
   Like racking engines!

The Poems and Songs of Robert Burns
Volume II
Address to the Toothache
Stanza I

Busch, Wilhelm

A toothache, not to be perverse,
Is an unmitigated curse . . .

German Satirical Writings
The Poet Thwarted (p. 161)

Carroll, Lewis

"Do I look very pale?" said Tweedledum, coming up to have his helmet tied on. (He *called* it a helmet, though it certainly looked much more like a saucepan.)

"Well—yes—a *little*," Alice replied gently.

"I'm very brave generally," he went on in a low voice: "only to-day I happen to have a headache."

"And *I've* got a toothache!" said Tweedledee, who had overheard the remark. "I'm far worse than you!"

The Complete Works of Lewis Carroll
Through the Looking Glass
Chapter 4 (p. 193)

Collins, John
Maria one Morning was smitten full sore,
With the Tooth-ach's unmerciful Pang;
And she vow'd, if she liv'd to the Age of Five-score,
That she still should remember the Fang: . . .

Scripscrapologia
Excuse for Oblivion, L. 1–4

Esar, Evan
[Toothache] The pain in a tooth that sometimes drives you to extraction.

Esar's Comic Dictionary

[Toothache] The torture where a person is in pain and the pain is in him.

Esar's Comic Dictionary

Fuller, Thomas
The tongue is ever turning to the aching tooth.

Gnomologia
Number 4796

Gilbert, Sir William Schwenck
Roll on, thou ball, roll on!
Through pathless realms of Space
      Roll on!
What though I'm in a sorry case?
What though I cannot meet my bills?
What though I suffer toothache's ills?
What though I swallow countless pills?

In Helen and Lewis Melville
An Anthology of Humorous Verse
To the Terrestrial Globe

Heath-Stubbs, John
Venerable Mother Toothache
Climb down from the white battlements,
Stop twisting in your yellow fingers
The fourfold rope of nerves.

Collected Poems 1943–1987
A Charm Against the Toothache

James, Henry
He might have been a fine young man with a bad toothache; with the first even of his life. What ailed him above all, she felt, was that trouble was new to him . . .

The Spoils of Poynton
Chapter 8

Josselyn, John
... for the Toothache I have found the following medicine very available, Brimstone and Gunpowder compounded with butter; rub the mandible with it, the outside being first warm'd.

Two Voyages to New-England
The Second Voyage (pp. 128–9)

Melville, Herman
Another has the toothache: the carpenter out pincers, and clapping one hand upon his bench bids him be seated there; but the poor fellow unmanageably winces under the unconcluded operation; whirling round the handle of his wooden vice, the carpenter signs him to clap his jaw in that, if he would have him draw the tooth.

Moby Dick
Chapter cvii

Proverb
The tooth-ache is more ease,
than to deale with ill people.

In George Herbert
Outlandish Proverbs
#558

Ray, John
Who hath aching teeth hath ill tenants.

A Complete Collection of English Proverbs (p. 26)

Shakespeare, William
What! sigh for the toothache?

Much Ado About Nothing
Act V, Scene I, L. 21

For there was never yet philosopher
That could endure the toothache patiently.

Much Ado About Nothing
Act V, Scene I, L. 35–36

Being troubled with a raging tooth,
I could not sleep.

Othello
Act III, Scene III, L. 414–415

He that sleeps feels not the toothache.

Cymbeline
Act V, Scene IV, L. 177

Shaw, George Bernard

The man with toothache thinks everyone happy whose teeth are sound.

The Revolutionist's Handbook & Pocket Companion
Maxims for Revolutionists
Greatness (p. 56)

Unknown

If in your teeth you hap to be tormented,
By meane some little wormes therein do breed,
Which paine (if heed be tane) may be prevented,
By keeping cleane your teeth, when as you feede:
Burne Francomsence (a gum not evil sented),
Put Henbane unto this, and Onyon seed,
And with a tunnel to the tooth that's hollow,
Convey the smoake thereof, and ease shall follow.

Notes and Queries
5th Series, Volume vi
July 29, 1826 (p. 97)

As Peter was sitting alone on a marble stone, Christ came to him and said: 'Peter what is the matter with you?'

'The toothache, my Lord God.'

'Arise, Peter, and be free'; And every man and woman will be cured of the toothache, who shall believe these words. I do this in the name of God.

The Academy
Toothache Charms (p. 258)
Volume 31, Number 779, April 9, 1887

Kookaburra sits on the old gum tree
With a toothache bad as can be
Ha kookaburra, ha kookaburra
Didn't save any for me.

Kookaburra

May he swell with the gout, may his grinders fall out,
May he roar, bawl and shout, with the horrid toothache.

Recorded by Tommy Makem
Irish Songs of Rebellion
Nell Flaherty's Drake

I had me a toothache; it hurt me so bad
I went to the dentist; now I'm sad
He yanked with his fingers and pulled with his thumb
He left all my teeth; but he pulled out my gums.

In Homer and Jethro
Homer and Jethro: Country Comedy
RCA 1971, CXS-9012(e)
So Long (It's Been Good to Know Yuh!)

Cough and the world coughs with you. Fart and you stand alone.
Trevor Griffiths – (See p. 56)

TRANQUILIZERS

Lerner, Max

What is dangerous about the tranquilizers is that whatever peace of mind they bring is a packaged peace of mind. Where you buy a pill and buy peace with it, you get conditioned to cheap solutions instead of deep ones.

The Unfinished Country
The Assault on the Mind
June 25, 1957 (p. 203)

Unknown

If everyone took tranquilizers, no one would need them.

In Evan Esar
20,000 Quips and Quotes

A tranquilizer is a pill for the body to treat an ill of the mind.

In Evan Esar
20,000 Quips and Quotes

TREATMENT

Crichton-Browne, Sir James
On your second visit you ask him if the medicine you had prescribed for him had done any good, and the answer is: "Well, it has done no harm," from which you may infer that it has been highly beneficial to him.

From the Doctor's Notebook
The Ungracious Patient (p. 200)

Latham, Peter Mere
It is an instructive fact that, as the knowledge of disease has increased, the practice of medicine has been less and less conversant with cures and more and more conversant with treatment.

In William B. Bean
Aphorisms from Latham (p. 60)

Treatment is concerned with the individual patient and leaves his disease to take care of itself.

In William B. Bean
Aphorisms from Latham (p. 60)

It is not possible that the treatment of diseases shall be ever set at rest by the consent of physicians, or that fixed and uniform plans and remedies shall ever be adopted in cases bearing the same nosological name and character.

In William B. Bean
Aphorisms from Latham (p. 61)

Nye, Bill
When you have spinal meningitis, however, the doctor tackles you with bromides, ergots, ammonia, iodine, chloral hydrate, codi, bromide of ammonia, hasheesh, bismith, valerianate of ammonia, morphine sulph., nux vomica, turpentine emulsion, vox humana, rex magnus, opium, cantharides, Dover's powders, and other bric-a-brac. These remedies

are masticated and acted upon by the salivary glands, passed down the esophagus, thrown into the society of old gastric, submitted to the peculiar motion of the stomach and thoroughly chymified, then forwarded through the pyloric orifice into the smaller intestine, where they are touched up with bile, and later on handed over through the lacteals, thoracic duct, etc., to the vast circulatory system. Here it is yanked back and forth through the heart, lungs and capillaries, and if anything is left to fork over to the disease, it has to squeeze into the long, bony, air-tight socket that holds the spinal cord.

Remarks
Spinal Meningitis (p. 123)

Unknown

Mrs. X had bosom trouble,
  She was flat across the bow;
Then she took three bottles of Compound,
  Now they milk her like a cow!
Oh we sing, we sing, we sing
  Of Lydia Pinkham, Pinkham, Pinkham
And her love for the human race.

In J.C. Furnas
The Life and Times of the Late Demon Rum (p. 183)

Unknown

I have an earache:

2000 B.C. —Here, eat this root.

1000 A.D. —That root is heathen. Here, say this prayer.

1850 A.D. —That prayer is superstition. Here, drink this potion.

1940 A.D. —That potion is snake oil. Here, swallow this pill.

1985 A.D. —That pill is ineffective. Here, take this antibiotic.

2000 A.D. —That antibiotic is artificial. Here, eat this root.

A Short History of Medicine

TRICHINOSIS

Bierce, Ambrose

TRICHINOSIS, *n*. The pig's reply to proponents of porcophagy.

Moses Mendelssohn having fallen ill sent for a Christian physician, who at once diagnosed the philosopher's disorder as trichinosis, but tactfully gave it another name. "You need an immediate change of diet," he said; "you must eat six ounces of pork every other day."

"Pork?" shrieked the patient—"pork? Nothing shall induce me to touch it!"

"Do you mean that?" the doctor gravely asked.

"I swear it!"

"Good!—then I will undertake to cure you."

The Enlarged Devil's Dictionary

TRUTH

Holmes, Oliver Wendell
Your patient has no more right to all the truth you know than he has to all the medicine in your saddlebags, if you carry that kind of cartridge-box for the ammunition that slays disease.

Medical Essays
The Young Practitioner (p. 388)

Mayo, William J.
Medical science aims at the truth and nothing but the truth.

Journal of the Indiana Medical Association
The Influence of Ignorance and Neglect on the
Incidence and Mortality of Cancer
Volume 17, 1924

Osler, Sir William
Start out with the conviction that absolute truth is hard to reach in matters relating to our fellow creatures, healthy or diseased, that slips in observation are inevitable even with the best trained facilities, that errors in judgment must occur in the practice of an art which consists largely in balancing probabilities;—start, I say, with this attitude of mind, and mistakes will be acknowledged and regretted; but instead of a slow process of self-deception, with ever increasing inability to recognize truth, you will draw from your errors the very lessons which may enable you to avoid their repetition.

Aequanimitas
Teacher and Student (p. 38)

Reichenbach, Hans
He who searches for truth must not appease his urge by giving himself up to the narcotic of belief.

In Ruth Renya
The Philosophy of Matter in the Atomic Era (p. 16)

TUBERCULOSIS

Bethune, Norman
There is a rich man's tuberculosis and a poor man's tuberculosis. The rich man recovers and the poor man dies . . .

Canadian Medical Association Journal
A Plea for Early Compression
July 1932

Bunyan, John
. . . the Captain of all the men of death that came against him to take him away, was the Consumption, for t'was that that brought him down to the grave.

The Life and Death of Mr. Badman (p. 239)

TUMOR

Abse, Dannie
I know the colour rose, and it is lovely,
But not when it ripens in a tumour;
And healing greens, leaves and grass, so springlike,
In limbs that fester are not springlike.

Collected Poems 1948–1976
Pathology of Colours (p. 89)

URINALYSIS

Armour, Richard

Some bring their sample in a jar,
　　Some bring it in a pot,
Some bring a sample hardly ample,
　　While others bring a lot.

The Medical Muse
Urinalysis

VACCINATION

Epps, John

. . . no discovery, beneficial to man, has ever been effected which was received at its promulgation with a candid welcome. The front of opposition has always been raised, and the boon-bestowing visitor has been tempted, by his rude reception, to leave such ungrateful creatures to themselves. Vaccination, therefore, cannot be supposed to be an exception.

The Life of John Walker
Chapter IX (pp. 312–13)

Huxley, Thomas Henry

If [my next-door neighbor] is allowed to let his children go unvaccinated, he might as well be allowed to leave strychnine lozenges about in the way of mine . . .

Critiques and Addresses
Administrative Nihilism (p. 10)

Unknown

Typhoid fever may be prevented by fascination.

In Alexander Abingdon
Bigger & Better Boners (p. 74)

If more people would get intoxicated much smallpox could be prevented.

In Alexander Abingdon
Bigger & Better Boners (p. 75)

VIRUS

Newman, Michael
Observe this virus: think how small
Its arsenal, and yet how loud its call;
It took my cell, now takes your cell,
And when it leaves will take our genes as well.

The Sciences
Cloned Poem
1982

Unknown
A virus is a Latin word translated by doctors to mean "Your guess is as good as mine."

Source unknown

X-RAYS

Davies, Robertson
My insides has always fascinated photographers, though none of them has ever shown the least enthusiasm about my outside. "Lock your hands behind your head, tie your legs in a knot, cross your eyes, touch the end of your nose with the tip of your tongue; now hold still . . ." says the X-ray technician . . .

The Table Talk of Samuel Marchbanks (p. 166)

Jauncey, G.E.M.
O Roentgen, then the news is true
  And not a trick of idle rumor
That bids us each beware of you
  And of your grim and graveyard humor.

Scientific American
February 22, 1896

Pasveer, B.
The X-ray images were trusted for their ability to represent reality, but in the pre-Röntgen era reality looked enormously different from the shadows that were now said to be mirroring the inner parts of patients.

Sociology of Health and Illness
Knowledge of Shadows: The Introduction of X-ray Images in Medicine
Volume 11, 1989 (p. 361)

Russell, Bertrand
Everybody knows something about X-rays, because of their use in medicine. Everybody knows that they can take a photograph of the skeleton of a living person, and show the exact position of a bullet lodged in the brain. But not everybody knows why this is so. The reason is that the capacity of ordinary matter for stopping the rays varies approximately as the fourth power of the atomic number of the elements concerned . . .

The ABC of Atoms (p. 106)

Russell, L.K.
She is so tall, so slender, and her bones—
Those frail phosphates, those carbonates of lime—
Are well produced by cathode rays sublime,
By oscillations, amperes and by ohms.
Her dorsal vertebrae are not concealed
By epidermis, but are well revealed.

Life
Lines on an X-ray Portrait of a Lady
March 12, 1896

Thomas, Gwyn
Her first big economic drive will be to replace X-ray by hearsay.

The Keep
Act Two, Scene One (p. 72)

Unknown
The Roentgen Rays, The Roentgen Rays
What is this craze,
The town's ablaze,
With the new phase
of X-rays' ways.
I'm full of daze,
Shock and amaze,
For nowadays,
I hear they'll gaze,
Thro' cloak and gown—and even stays,
These naughty, naughty Roentgen Rays.

In John G. Taylor
The New Physics (p. 46)

I hope you will examine the X-ray films less and me more.

Source unknown

BIBLIOGRAPHY

Abingdon, Alexander. *Bigger & Better Boners*. The Viking Press, New York. 1952.

Abse, Dannie. *Collected Poems 1948–1976*. University of Pittsburgh Press, Pittsburgh. 1977.

Ace, Goodman. *The Fine Art Of Hypochondria*. Doubleday & Company, Inc., Garden City. 1966.

Adams, Cedric. *Poor Cedric's Almanac*. Doubleday & Company, Inc., Garden City. 1952.

Adams, Henry. *The Education of Henry Adams*. The Modern Library, New York. 1918.

Adams, W. Davenport. *English Epigrams*. George Routledge and Sons, London. No date.

Akenside, Mark. *The Pleasures of the Imagination*. Associated University Press, London. 1996.

Amiel, Henri-Frédéric. *Amiel's Journal*. Translated by Mrs. Humphrey Ward. A.L. Burt Company. 1885.

Anderson, Peggy. *Nurse*. Berkley Books, New York. 1978.

Arderne, John. *Treatises of Fistula in Ano*. Kegan Paul, Trench, Trubner & Co., Ltd., London. 1910.

Aristotle. *Categories* in *Great Books of the Western World*. Volume 8. Encyclopaedia Britannica, Inc., Chicago. 1952.

Aristotle. *Politics* in *Great Books of the Western World*. Volume 9. Encyclopaedia Britannica, Inc., Chicago. 1952.

Aristotle. *Posterior Analytics* in *Great Books of the Western World*. Volume 8. Encyclopaedia Britannica, Inc., Chicago. 1952.

Armour, Richard. 'Taking No Chances' in *Quote, The Weekly Digest*. October 1, 1967.

Armour, Richard. *The Medical Muse*. McGraw-Hill Book Company, Inc., New York. 1963.

Armstrong, John. *Art of Preserving Health*. Printed for A. Miller, London. 1747.

Arnauld, Antoine. *The Art of Thinking: Port-Royal Logic*. The Bobbs-Merrill Co., Indianapolis. 1964.

Arnold, Matthew. *Poems by Matthew Arnold*. Volume Two. Macmillan and Co., London. 1881.

Atkinson, Caroline P. *Letters of Susan Hale*. Marshall Jones Company, Boston. 1919.

Aurelius, Marcus. *Meditations* in *Great Books of the Western World*. Volume 12. Encyclopaedia Britannica, Inc., Chicago. 1952.

Aylett, Robert. *Peace with Her Foure Gardners*. Printed for John Teage, London. 1622.

Bacon, Francis. *Advancement of Learning* in *Great Books of the Western World*. Volume 30. Encyclopaedia Britannica, Inc., Chicago. 1952.

Bacon, Francis. *Essays, Advancement of Learning, New Atlantis, and Other Pieces*. The Odyssey Press, Inc., New York. 1937.

Bahya, ben Joseph ibn Paauda. *Duties of the Heart*. Bloch Publishing Company, New York. 1941.

Baldwin, Joseph G. *The Flush Times of Alabama and Mississippi*. Louisiana State University Press, Baton Rouge. 1987.

Balzac, Honoré de. *The Physiology of Marriage*. Liveright Publishing Corporation, New York. 1932.

Baring, Maurice. *The Black Prince and Other Poems*. John Lane, London. 1903.

Barnard, Christian N. 'People' in *Time*. October 31, 1969.

Barnes, Djuna. *Nightwood*. Harcourt, Brace & Company, New York. 1937.

Barring-Gould, William S. *The Lure of the Limerick*. Clarkson N. Potter, Inc., New York. 1967.

Barrow, John D. *The Artful Universe*. Clarendon Press, Oxford. 1995.

Barrow, John D. *Impossibility*. Oxford University Press, Oxford. 1998.

Barss, Peter. 'Injuries Due to Falling Coconuts' in *The Journal of Trauma*. Volume 24, Number 11. 1984.

Baum, Harold. *The Biochemists' Handbook*. Pergamon Press, Oxford. 1982.

Bayliss, William Maddock. *Principles of General Physiology*. Longmans, Green and Co., London. 1920.

Beacock, Cal. 'The Pundit' in *Reader's* Digest. January 1986.

Bean, William B. *Aphorisms from Latham*. The Prairie Press, Iowa City. 1962.

Beard, George M. 'Experiments with Living Human Beings' in *Popular Science Monthly*. Volume 14. 1879.

Beaver, Wilfred. *Quote, The Weekly Digest*. May 19, 1968.

Beckett, Samuel. *All that Fall*. Grove Press, Inc., New York. 1957.

Beckett, Samuel. *Murphy*. Grover Press, Inc., New York. 1957.

Bell, Eric T. *Mathematics: Queen and Servant of Science*. McGraw-Hill Book Co., Inc., New York. 1940.

Bell, Robert. *The Poems of Robert Greene, Christopher Marlowe, and Ben Jonson*. Hurst and Company, New York. No date.

Belloc, Hilaire. *Cautionary Tales for Children*. Duckworth, London. No date.

Benchley, Robert. *Benchley or Else*. Harper & Brothers, New York. 1947.

Benjamin, Arthur. 'A Free Bike with Your Braces' in *Newsweek*. May 5, 1986.

Bernard, Christiaan. 'People' in *Time*. October 31, 1969.

Bernard, Claude. *An Introduction to the Study of Experimental Medicine*. Henry Schuman, Inc. 1949.

Bernstein, Al. *Quote, The Weekly Digest*. July 28, 1968.

Bethune, Norman. 'A Plea for Early Compression' in *Canadian Medical Association Journal*. July 1932.

Bierce, Ambrose. *The Enlarged Devil's Dictionary*. Doubleday & Company, Inc., Garden City. 1967.

Bloom, Samuel W. *The Doctor and His Patient*. Russell Sage Foundation, New York. 1963.

Bond, John and Bond, Senga. *Sociology and Health Care*. Churchill Livingstone, Edinburgh. 1986.

Boorde, Andrew. *The Wisdom of Andrew Boorde*. E. Bacus, Leicester. 1936.

Booth, Gothard. *The Cancer Epidemic: Shadow of the Conquest of Nature*. The Edwin Mellen Press, New York. 1974.

Boulding, Kenneth E. *The Image*. The University of Michigan Press, Ann Arbor. 1956.

Bowen, Elizabeth. *The Death of the Heart*. Jonathan Cape, London. 1938.

Brackenridge, Hugh Henry. *Modern Chivalry*. American Book Company, New York. 1937.

Bradford, Maynerd. *Quote, The Weekly Digest*. September 8, 1968.

Brooks, Paul. *The House of Life: Rachel Carson at Work*. Houghton Mifflin Company, Boston. 1972.

Browne, Sir Thomas. *Religio Medici*. At the University Press, Cambridge. 1955.

Browne, Sir Thomas. *The Works of Sir Thomas Browne*. Volume One. John Grant, Edinburgh. 1927.

Browne, Sir Thomas. *The Works of Sir Thomas Browne*. Volume Three. John Grant, Edinburgh. 1927.

Browning, Elizabeth Barrett. *Aurora Leigh*. The Walter Scott Publishing Co., Ltd., London. No date.

Buchan, William. *Domestic Medicine*. Twentieth Edition. Waterford, New York. 1797.

Bunyan, John. *The Life and Death of Mr. Badman*. Oxford University Press, London. 1929.

Burgess, Anthony. *Nothing Like the Sun*. W.W. Norton & Company, Inc., New York. 1964.

Burns, Olive Ann. *Cold Sassy Tree*. Ticknor & Fields, New York. 1984.

Burns, Robert. *The Poems and Songs of Robert Burns*. Volume II. At the Clarendon Press, Oxford. 1968.

Burton, Robert. *The Anatomy of Melancholy*. Clarendon Press, Oxford. 1990.

Busch, Wilhelm. *German Satirical Writings*. The Continuum Publishing Company, New York. 1984.

Butler, Samuel. *Erewhon*. E.P. Dutton & Company, New York. 1917.

Butler, Samuel. *Hudibras*. At the Clarendon Press, Oxford. 1967.

Butler, Samuel. *Samuel Butler's Notebooks*. Jonathan Cape, London. 1951.

Butler, Samuel. *The Poetical Works of Samuel Butler*. Volume I. James Nichol, Edinburgh. 1854

Butler, Samuel. *The Poetical Works of Samuel Butler*. Volume II. James Nichol, Edinburgh. 1854

Butler, Samuel. *The Way of All Flesh*. E.P. Dutton & Company, New York. 1914.

Byrne, Josefa Heifetz. *Mrs. Byrne's Dictionary*. Citadel Press, Secaucus. 1974.

Byron, Lord George Gordon. *Don Juan*. Blue Ribbon Books, New York. 1932.

Byron, Lord George Gordon. *The Poetical Works of Lord Byron*. William Collins, Sons & Co., Limited, London. No date.

Cairns, Sir Hugh. 'The Student's Objective' in *The Lancet*. Volume 257. October 8, 1949.

Camden, William. *Remains Concerning Britain*. John Russell Smith, London. 1870.

Cardozo, Benjamin N. *The Paradoxes of Legal Science*. Columbia University Press, New York. 1928.

Carlyle, Thomas. *Characteristics*. No information on bibliography.

Carroll, Lewis. *The Complete Works of Lewis Carroll*. The Modern Library, New York. No date.

Causley, Charles. *Collected Poems*. Macmillan, London. 1992.

Cecil, Russell L. and Leob, Robert F. *Textbook of Medicine*. 9th edition. Saunders, Philadelphia. 1954.

Cervantes, Miguel de. *Don Quixote de la Mancha* in *Great Books of the Western World*. Volume 29. Encyclopaedia Britannica, Inc., Chicago. 1952.

Chatton, Milton J. *Quotable Quotes*. March 13, 1966.

Chatton, Milton J. 'Methanosis' in *The New England Journal of Medicine*. Volume 287, Number 2. 1972.

Chekhov, Anton. *Letters on the Short Story*. Benjamin Blom, Publisher, New York. 1964.

Chekhov, Anton. *Note-Book of Anton Chekhov*. B.W. Huebsch, Inc., New York. 1922.

Chekhov, Anton. *The Cherry Orchard*. English version by Sir John Gielgud. Heinemann, London. 1963.

Chekhov, Anton. *The Portable Chekhov*. The Viking Press, New York. 1947.

Chesterfield, Philip Dormer Stanhope. *The Letters of Philip Dormer Stanhope*. Volume Five. Eyre and Spottiswoode Ltd., London. 1932.

Chesterton, G.K. *Come to Think of It*. Methuen & Co., Ltd., London. 1930.

Chesterton, G.K. *Generally Speaking*. Dodd Mean & Co., New York. 1929.

Christie, Agatha. *At Bertram's Hotel*. The Crime Club, London. 1965.

Christie, Agatha. *Endless Night*. Collins, London. 1967.

Christy, Robert. *Proverbs, Maxims and Phrases of All Ages*. G.P. Putnam's Sons, New York. 1888.

Clemens, Clara. *My Father Mark Twain*. Harper & Brothers Publishers, New York. 1931.

Clendening, Logan. *Modern Methods of Treatment*. The C.V. Mosby Company, St. Louis. 1924.

Clode, Edward J. *Jokes for All Occasions*. Edward J. Clode, New York. 1921.

Clowes, William. *Selected Writings of William Clowes*. Harvey & Blythe Ltd., London. 1948.

Coates, Florence Earle. *Poems*. Volume Two. Houghton Mifflin Company, Boston. 1916.

Collingwood, R.G. *The New Leviathan*. At the Clarendon Press, Oxford. 1942.

Collins, John. *Scripscrapologia*. Published by John Collins, Birmingham. 1804.

Colman, George. *Broad Grins*. Garland Publishing, Inc., New York. 1977.

Colton, Charles C. *Lacon*. William Gowans, New York. 1849.

Compton-Burnett, I. *A Family and A Fortune*. Eyre & Spottiswoode, London. 1949.

Coope, Robert. *The Quiet Art*. E. & S. Livingstone, Ltd., Edinburgh. 1952.

Cordus, Euricus. 'The Three Characters of a Physician' in *Annals of Medical History*. Volume I. 1917.

Cowper, William. *The Task*. Lewis and Sampson, Boston. 1842.

Crabbe, George. *Tales in Verse*. Printed for J. Hatchard, Bookseller to Her Majesty, London. 1812.

Crabbe, George. *The Poetical Works of George Crabbe*. Grigg & Eliot, Philadelphia. 1847.

Crane, Edward V. *American Engineer*. Volume 26, No. 8. August 1956.

Crichton-Browne, Sir James. *From the Doctor's Notebook*. Duckworth, London. 1937.

Crichton-Browne, Sir James. *The Doctor Remembers*. Duckworth, London. 1938.

Crichton-Browne, Sir James. *The Doctor's After Thoughts*. Ernest Benn Limited, London. 1932.

Croll, Oswald. *Philosophy Reformed and Improved in Four Profound Tractates*. Printed by M.S. for Lodwick Lloyd, London. 1657.

Cross, Hardy. *Engineers and Ivory Towers*. McGraw-Hill Book Company, Inc., New York. 1952.

Crothers, Samuel McChord. *The Gentle Reader*. Houghton Mifflin Company, Boston. 1903.

Cumston, C.G. *An Introduction to the History of Medicine from the Time of the Pharaohs to the End of the XVIIIth Century.* Kegan Paul, Trench Trubner and Co., London. 1926.

Cushing, Harvey. *The Life of Sir William Osler.* Volume I. At the Clarendon Press, Oxford. 1925.

Cushing, Harvey. *The Life of Sir William Osler.* Volume II. At the Clarendon Press, Oxford. 1925.

Cvikota, Raymond J. *Quote, The Weekly Digest.* April 7, 1968.

Cvikota, Raymond J. *Quote, The Weekly Digest.* June 9, 1968.

Cvikota, Raymond J. *Quote, The Weekly Digest.* October 27, 1968.

Czarnomski, F.B. *The Wisdom of Winston Churchill.* George Allen and Unwin Ltd., London. 1956.

da Costa, J. Chalmers. *The Trials and Triumphs of the Surgeon.* Dorrance & Company, Philadelphia. 1944.

da Vinci, Leonardo. *Leonardo da Vinci's Notebooks.* Duckworth & Co., London. 1906.

Dagi, Teodoro Forcht. 'Anatomy of the Brain and Spinal Medulla: A Manual for Students' in *The New England Journal of Medicine.* Volume 286, Number 18. May 4, 1972.

Davies, P.C.W. and Julian Brown. *Superstrings: A Theory of Everything?* Cambridge University Press, Cambridge. 1988.

Davies, Robertson. *The Table Talk of Samuel Marchbanks.* Clarke, Irwin & Company Limited, Toronto. 1949.

Davis, Adelle. *Let's Eat Right to Keep Fit.* Harcourt, Brace & World, Inc., New York. 1954.

Day, Clarence. *This Simian World.* Alfred A. Knopf, New York. 1941.

De Bakey, Michael E. 'Heart of the Matter' in *Newsweek.* June 6, 1966.

de Guevara, Antonio. *The Familiar Epistles of Sir Anthonie of Guevara.* Imprinted for Ralph Newberrie, London. 1577.

de La Fontaine, Jean. *La Fontaine: Selected Fables.* Translated by James Michie. The Viking Press, New York. 1979.

de Madariaga, Salvador. *Essays with a Purpose.* Hollis & Carter, London. 1954.

Defoe, Daniel. *Selected Poetry and Prose of Daniel Defoe.* Holt, Rinehart and Winston, New York. 1968.

Dennis, F.S. *The History and Development of Surgery during the Past Century.* Philadelphia. 1905.

Di Bacco, Babs Z. 'Leisure Gap' in *American Journal of Nursing.* January 1969.

Dickens, Charles. *A Tale of Two Cities.* Thomas Y. Cramell Company, New York. 1904.

Dickens, Charles. *The Pickwick Papers.* Volume II in *The Works of Charles Dickens.* The University Society, New York. 1908.

Dickens, Charles. *David Copperfield*. Volume II in *The Works of Charles Dickens*. The University Society, New York. 1908.

Dickinson, Emily. *The Complete Poems of Emily Dickinson*. Little Brown and Company, Boston. 1960.

Dickinson, Emily. *Poems (1890–1896)*. Scholars' Facsimiles & Reprints, Gainsville. 1967.

Dickinson, Richard W. *Transactions of the Association of American Physicians*. Volume 75, Number 1. 1962.

Dimmick, Edgar L. 'The In-Side of Two' in *Journal of the American Medical Association*. Volume 199, Number 6. 1967.

Donaldson, T.B. *An Apropos Alphabet*. W.S. Sterling & Co., New York. 1900.

Donleavy, J.P. *The Ginger Man*. Dell Publishing Co., Inc., New York. 1965.

Donne, John. *An Anatomy of the World*. The John Hopkins Press, Baltimore. 1963.

Donne, John. *Devotions Upon Emergent Occasions*. McGill-Queen's University Press, Montreal. 1975.

Donne, John. *The Poems of John Donne*. Oxford University Press, London. 1937.

Drake, Daniel. *An Introductory Lecture, on the Means of Promoting the Intellectual Improvement of the Students*. 2nd Edition. 1844.

Drake, Daniel. *Practical Essays on Medical Education, and the Medical Profession*. Roff & Young, Cincinnati. 1832.

Drake, Daniel. *Western Journal of Medicine and Surgery*. N.S. II: 355. October 1844.

Dryden, John. *The Poems of John Dryden*. Volume I. At the Clarendon Press, Oxford. 1953.

Dryden, John. *The Poems of John Dryden*. Volume IV. At the Clarendon Press, Oxford. 1953.

Dryden, John. *The Poetical Works of John Dryden*. Volume V. Bell and Dald, London. No date.

Du Bartas, Guillaume de Saluste. *Du Bartas and His Divine Weekes and Works*. Printed by Robert Young, London. 1641.

Dubos, René. *Man Adapting*. Yale University Press, New Haven. 1965.

Dubos, René. *Mirage of Health*. Harper & Brothers Publishers, New York. 1959.

Dubos, René and Escande, Jean-Paul. *Quest: Reflections on Medicine, Science, and Humanity*. Translated by Patricia Runum. Harcourt Brace Jovanovich, New York. 1980.

Duffy, John C. and Litin, Edward M. 'Psychiatric Morbidity of Physicians' in *Journal of the American Medical Association*. Volume 189. 1964.

Dunlap, William. *Recollections of the War of 1812*. Historical Publishing Co., Toronto. 1908.

Dunlap, William. *The Memoirs of a Water Drinker*. Volume I. Bancroft and Holley, New York. 1836.

Dunne, Finley Peter. *Mr. Dooley's Opinions*. Harper & Brothers Publishers, New York. 1906.

Dunne, Finley Peter. *Mr. Dooley: On Making a Will and Other Necessary Evils*. Charles Scribner's Sons, New York. 1919.

Dunne, Finley Peter. *Mr. Dooley Says*. Charles Scribner's Sons, New York. 1910.

Eddy, Mary Baker. *Science and Health with Key to the Scriptures*. Trustees under the Will of Mary Baker G. Eddy, Boston. 1906.

Editor, The Louisville Journal. *Prenticeana*. Derby & Jackson, New York. 1860.

Ehrensvärd, Gösta Carl Henrik. *Man on Another World*. Translated by Lennart and Kajsa Rodem. The University of Chicago Press, Chicago. 1965.

Eisenschiml, Otto. *The Art of Worldly Wisdom*. Duell, Sloan and Pearce, New York. 1947.

Eldridge, Paul. *Maxims for a Modern Man*. Thomas Yoseloff, New York. 1965.

Eliot, T.S. *Four Quartets*. Harcourt, Brace and Company, New York. 1943.

Eliot, T.S. *The Cocktail Party*. Harcourt, Brace and Company, New York. 1950.

Eliot, T.S. 'Whispers of Immortality' in *The Complete Poems and Plays 1909–1950*. Harcourt, Brace and Company, New York. 1952.

Eliot, T.S. *The Elder Statesman*. Farrar, Straus and Cudahy, New York. 1959.

Emerson, Ralph Waldo. *Complete Works*. Volume 6. Houghton Mifflin Company, Boston. 1904.

Emerson, Ralph Waldo. *Journals*. Volume V. Houghton Mifflin Company, Boston. 1913.

Emerson, Ralph Waldo. *Journals*. Volume IX. Houghton Mifflin Company, Boston. 1913.

Emerson, Ralph Waldo. *Natural History of Intellect*. Houghton Mifflin and Company, Boston. 1893.

Epps, John. *The Life of John Walker, M.D.* Whittaker, Treacher, and Co., London. 1831.

Esar, Evan. *20,000 Quips & Quotes*. Barnes & Noble Books, New York. 1968.

Esar, Evan. *Esar's Comic Dictionary*. Fourth Edition. Doubleday & Company, Inc., Garden City. 1983.

Euripides. *The Plays of Euripides* in *Great Books of the Western World*. Volume 5. Encyclopaedia Britannica, Inc., Chicago. 1952.

Fabing, Howard and Marr, Ray. *Fischerisms*. The Science Press Printing Co., Lancaster. 1937.

Farris, Jean. *Quote, The Weekly Digest*. February 18, 1968.

Farris, Jean. *Quote, The Weekly Digest*. August 4, 1968.

Faulkner, William. *The Sound and the Fury*. Random House, New York. 1984.

Fechner, Gustav. *Life After Death*. Pantheon Books, New York. 1943.

Field, Eugene. *The Poems of Eugene Field*. Charles Scribner's Sons, New York. 1941.

Fielding, Henry. *Tom Jones* in *Great Books of the Western World*. Volume 37. Encyclopaedia Britannica, Inc., Chicago. 1952.

Fillery, Frank. *Quote, The Weekly Digest*. February 12, 1967.

Fillery, Frank. *Quote, The Weekly Digest*. November 12, 1967.

Fixx, James. *The Complete Book of Running*. Random House, New York. 1977.

Flaubert, Gustave. *Dictionary of Accepted Ideas*. Max Reinhardt, London. 1954.

Flexner, Abraham. 'Is Social Work a Profession?' in *School and Society*. Volume 1. 1915.

Florio. *Firste Fruites* in *Memoirs of the Faculty of Literature and Politics, Taihoku Imperial University*. Volume III, Number 1. Published by the Taihoku Imperial University, Formosa. July 1936.

Ford, John. *The Lovers Melancholy*. Da Capo Press, Amsterdam. 1970.

Foster, Nellis B. *The Examination of Patients*. W.B. Saunders Company, Philadelphia. 1923.

Fowke, Edith. *The Penguin Book of Canadian Folk Songs*. Penguin Books Canada, Markham. 1986.

Fox, Sir Theodore. 'Purposes of Medicine' in *The Lancet*. Volume 2. October 23, 1965.

Frank, Julia Bess. 'Dermatology' in *The New England Journal of Medicine*. Volume 297, Number 12. 1977.

Friedman, Shelby. *Quote, The Weekly Digest*. February 18, 1968.

Friedman, Shelby. *Quote, The Weekly Digest*. February 25, 1968.

Friedman, Shelby. *Quote, The Weekly Digest*. March 10, 1968.

Friedman, Shelby. *Quote, The Weekly Digest*. May 7, 1967.

Frisch, Otto. *What Little I Remember*. Cambridge University Press, Cambridge. 1979.

Fuller, Thomas. *Gnomologia*. Printed by S. Powell, Dublin. 1733.

Fuller, Thomas. *The Holy State*. Printed by R.D., Cambridge. 1642.

Furnas, J.C. *The Life and Times of the Late Demon Rum*. G.P. Putnam's Sons, New York. 1965.

Gabor, Dennis. *Inventing the Future*. Secker & Warburg, London. 1963.

Garth, Samuel. *Garth's Dispensary* I. Teil: Text. Inaugural-Dissertation zur Erlangung der philosophischen Doktorwürde der hohen philosophischen Fakultät der Grossherzoglich Badischen Ruprecht-Karls-Universität Heidelberg/Vorgelegt von Wilhelm Josef Leicht. Heidelberg. 1905.

Gay, John. *Fables I*. University of California, Los Angeles. 1967.

Ghalioungui, Paul. *Magic and Medical Science in Ancient Egypt*. Barnes & Noble, Inc., New York. 1965.

Gibran, Kahlil. *The Prophet*. Alfred A. Knopf, New York. 1964.

Gide, André. *Journals*. Volume 3. Americ-edit, Rio de Janeiro. 1943.

Gisborne, Thomas. *An Enquiry into the Duties of Men*. Printed for B. and J. White, London. 1744.

Glasow, Arnold. *Quote, The Weekly Digest*. June 18, 1967.

Glasow, Arnold. *Quote, The Weekly Digest*. August 27, 1967.

Glasser, Allen. *Quote, The Weekly Digest*. May 7, 1967.

Goethe, Johann Wolfgang von. *Faust*. Jonathan Cape & Harrison Smith, New York. 1930.

Goldenweiser, Alexander. *Robots or Gods*. Alfred A. Knopf, New York. 1931.

Good, Irving John. *The Scientist Speculates*. Basic Books, Inc., Publishers, New York. 1962.

Goodman, Ace. *The Fine Art of Hypochondria*. Doubleday & Company, Inc., Garden City. 1966.

Gordon, Richard. 'The Common Cold' in *Atlantic Monthly*. January 1955.

Gould, Donald. *The Black and White Medicine Show*. Hamish Hamilton, London. 1985.

Grant, Claud. *Quote, The Weekly Digest*. June 2, 1968.

Green, Celia. *The Decline and Fall of Science*. Hamilton, London. 1976.

Griffiths, Trevor. *The Comedians*. Faber and Faber, London. 1979.

Guest, Edward A. *Collected Verse of Edward A. Guest*. Reilly & Lee Co., Chicago. 1946.

Gunther, John. *Taken at the Flood: The Story of Albert D. Lasker*. Harper & Brothers Publishers, New York. 1960.

Haggard, Howard W. *Mystery, Magic and Medicine*. Doubleday, Doran & Company, Inc., Garden City. 1933.

Haldane, J.B.S. 'Cancer's a Funny Thing' in *New Statesman*. February 21, 1964.

Haliburton, Thomas C. *Sam Slick's Wise Saws and Modern Instances*. Hurst and Blackett, Publishers, London. 1853.

Hanson, Elayne Clipper. 'Paradox' in *American Journal of Nursing*. March 1969.

Hardy, Thomas. *The Dynasts*. The Macmillan Company, New York. 1931.

Hare, Hobart A. *Practical Diagnosis*. Lea Brothers & Co., Philadelphia. 1902.

Hargrove, Cecilia. 'Prayer of a Pediatric Night Nurse' in *American Journal of Nursing*. March 1968.

Harrington, Sir John. *Metamorphosis of Ajax*. A critical annotated edition by Elizabeth Story Donno. Columbia University Press, New York. 1962.

Harris, Joel Chandler. *Nights with Uncle Remus*. McKinley, Stone & MacKenzie, New York. 1911.

Harrison, Jane. *Reminiscences of a Student's Life*. L. and V. Woolf, London. 1925.

Harvey, Gabriel. *Works*. Volume I. Printed for Private Circulation Only, London. 1884.

Harvey, William. *An Anatomical Disquisition on the Motion of the Heart and Blood in Animals* in *Great Books of the Western World*. Volume 28. Encyclopaedia Britannica, Inc., Chicago. 1952.

Hawthorne, Nathaniel. *Mosses from an Old Manse: The Procession of Life*. David McKay, Publisher, Philadelphia. 1890.

Hayes, Heather. *Quote, The Weekly Digest*. February 25, 1968.

Hazlitt, W. Carew. *Shakespeare Jest Books*. Volume III. Willis & Sotheran, London. 1864.

Heaney, Robert P. 'The Calcium Craze' in *Newsweek*. January 27, 1986.

Heath-Stubbs, John. *Collected Poems 1943–1987*. Carcanet Press Limited, Manchester. 1988.

Heber, Reginald. *The Poetical Work of Bishop Heber*. Frederick Warne and Co., London. 1897.

Heller, Joseph. *Catch-22*. Dell Publishing, New York. 1989.

Hellerstein, Herman. 'Tests to Avoid Attack' in *Newsweek*. August 6, 1984.

Helmuth, William Tod. *Scratches of a Surgeon*. Wm. A. Chatterton and Company, Chicago. 1879.

Henry, Sam. *Songs of the People*. University of Georgia Press, Athens. 1990.

Heraclitus. *On Nature*. N. Murray, Baltimore. 1889.

Herbert, George. *Outlandish Proverbs*. Printed by T.P. for Humphrey Blunden. 1640.

Herodotus. *The History of Herodotus* in *Great Books of the Western World*. Volume 6. Encyclopaedia Britannica, Inc., Chicago. 1952.

Herold, Don. *The Happy Hypochondriac*. Dodd, Mead & Company, New York. 1962.

Herold, J. Christopher. *The Mind of Napoleon*. Columbia University Press, New York. 1955.

Herrick, Robert. *The Complete Poems of Robert Herrick*. Chatto and Windus, London. 1876.

Heschel, Abraham J. *The Insecurity of Freedom*. Farrar, Straus & Giroux, New York. 1966.

Hewitt, Barnard. *The Doctor in Spite of Himself*. Row, Peterson & Company, Evanston. 1941.

Hippocrates. *Aphorisms* in *Great Books of the Western World*. Volume 10. Encyclopaedia Britannica, Inc., Chicago. 1952.

Hippocrates. *Laws* in *Great Books of the Western World*. Volume 10. Encyclopaedia Britannica, Inc., Chicago. 1952.

Hirsch, Nathaniel D. *Genius and Creative Intelligence*. Sci-Art Publishers, Cambridge. 1931.

Hoffmann, Friedrich. *Fundamenta Medicianae*. Translated by Lester S. King. Macdonald, London. 1971.

Hoffmann, Roald. *Gaps and Verges*. University of Central Florida Press, Orlando. 1990.

Holmes, Oliver Wendell. *The Guardian Angel*. Houghton Mifflin Company, Boston. 1867.

Holmes, Oliver Wendell. *Medical Essays*. Houghton Mifflin Company, Boston. 1911.

Holmes, Oliver Wendell. *Over the Teacups*. Houghton, Mifflin and Company, Boston. 1891.

Holmes, Oliver Wendell. *The Complete Poetical Works of Oliver Wendell Holmes*. Houghton Mifflin Company, Boston. 1908.

Holmes, Oliver Wendell. *The Professor at the Breakfast Table*. A.L. Burt Company, Publishers, New York. No date.

Hood, Thomas. *The Poetical Works of Thomas Hood*. Volume 1. William Collins, Sons, & Co., Limited, London. No date.

Hood, Thomas. *The Poetical Works of Thomas Hood*. Volume 2. William Collins, Sons, & Co., Limited, London. No date.

Hooker, Worthington. *Lessons from the History of Medical Delusions*. Baker & Scribner, New York. 1850.

Howard, Sidney. *Famous American Plays of the 1920s and the 1930s*. Selected by Kenneth MacCowan. The Fireside Theatre, Garden City. 1988.

Howe, Louise Kapp. *Moments on Maple Street*. Macmillan Publishing Company, New York. 1984.

Howell, James. *Proverbs*. Printed by J.G., London. 1659.

Howells, William Dean. *The Rise of Silas Lapham*. Printed for Members of The Limited Edition Club. 1961.

Hubbard, Elbert. *The Roycroft Dictionary*. The Roycrofters, East Aurora. 1914.

Hubbard, Kin. *Abe Martin's Barbed Wire*. The Bobbs-Merrill Company, Indianapolis. 1928.

Hubbard, Kin. *Abe Martin: Hoss Sense and Nonsense*. The Bobbs-Merrill Company, Indianapolis. 1926.

Hugill, Stan. *Shanties from the Seven Seas*. Routledge & K. Paul, New York. 1961.

Hulme, Keri. *The Bone People*. Louisiana State University Press, Baton Rouge. 1983.

Husserl, Edmund. *The Crisis of European Sciences and Transcendental Phenomenology*. Northwestern University Press, Evanston. 1970.

Huxley, Aldous. *Jesting Pilate*. Chatto & Windus, London. 1957.

Huxley, Aldous. *Proper Studies*. Chatto & Windus, London. 1927.

Huxley, Aldous. *Time Must Have a Stop*. Harper & Brothers Publishers, New York. 1944.

Huxley, Thomas H. *Critiques and Addresses*. Books for Libraries Press, Freeport. 1972.

Iannelli, Richard. *The Devil's New Dictionary*. Citadel Press, Secaucus. 1983.

Inge, William Ralph. *Outspoken Essays*. Second series. Longmans, Green and Co., New York. 1922.

Jackson, James. *Letters to a Young Physician*. Phillips, Samson and Company, Boston. 1855.

James, Alice. *The Diary of Alice James*. Dodd, Mead & Company, New York. 1964.

James, Henry. *The Spoils of Poynton*. The New Classics Series, Norfolk. 1924.

James, Henry. *Washington Square*. Thomas Y. Crowell Company, New York. 1970.

James, William. *Collected Essays and Reviews*. Longmans, Green and Co., New York. 1920.

Janssen, Johannes. *History of the German People at the Close of the Middle Ages*. Volume XIV. Translated by A.M. Christie. Kegan Paul, Trench, Trubner & Co. Ltd., London. 1909.

Jefferson, Thomas. *Inaugural Addresses of the Presidents of the United States*. U.S. Government Printing Office, Washington D.C. 1969.

Jerome, Jerome K. *Three Men in a Boat*. Time Life Books, New York. 1964.

Jewett, Sarah O. *Deephaven*. Houghton Mifflin Company, Boston. 1919.

Jhabvala, Ruth Prawer. *Travelers*. Harper & Row, Publishers, New York. 1973.

Johnson, Ernest. 'That Aching Back' in *Time*. July 14, 1980.

Johnson, Samuel. *Irene*. Printed for R. Dodsley, London. 1754.

Johnson, Samuel. *The Rambler*. Garland Publishing, Inc., New York. 1978.

Johnston, Alva. *The Legendary Mizners*. Farrar, Strauss & Young, New York. 1953.

Jonsen, Albert R. 'Watching the Doctor' in *New England Journal of Medicine*. Volume 308, Number 25. June 23, 1983.

Jonson, Ben. *Bartholomew Fair*. Yale University Press, New Haven. 1963.

Jonson, Ben. *Volpone*. Manchester University Press, Manchester. 1983.

Josselyn, John. *Two Voyages to New-England*. University Press of New England, Hanover. 1988.

Junior, Democritus. *The Anatomy of Melancholy*. Volume II. London. 1813.

Kafka, Franz. *The Complete Stories*. Schocken Books, New York. 1971.

Kafka, Franz. *Letters to Milena*. Translated by Tania and James Stern. Secker & Warbug, London. 1953.

Karch, Carroll S. *Quote, The Weekly Digest*. September 15, 1968.

Karch, Carroll S. *Quote, The Weekly Digest*. October 27, 1968.

Karmel, Marjorie. *Thank You, Dr. Lamaze*. J.B. Lippincott Company, Philadelphia. 1959.

Keats, John. *The Poems of John Keats*. At the University Printing House, Cambridge. 1966.

Kernan, F.C. *Quote, The Weekly Digest*. March 19, 1967.

Kerr, Jean. *Please Don't Eat the Daises*. Doubleday & Company, Inc., Garden City. 1954.

King, Alexander. *Rich Man, Poor Man, Freud and Fruit*. Simon and Schuster, New York. 1965.

Kipling, Rudyard. *A Book of Words*. Doubleday, Doran & Company, Inc., Garden City. 1928.

Kipling, Rudyard. *Collected Verse of Rudyard Kipling*. Doubleday, Page & Company, Garden City. 1925.

Kipling, Rudyard. *Just So Stories*. International Collectors Library, Garden City. 1912.

Kipling, Rudyard. *Kim*. Macmillan and Company, Ltd., London. 1900.

Kipling, Rudyard. *Mandalay*. Doxy, San Francisco. 1899.

Koestler, Arthur. *The Act of Creation*. The Macmillan Company, New York. 1964.

Kraus, Jack. *Quote, The Weekly Digest*. August 14, 1966.

Kraus, Jack. *Quote, The Weekly Digest*. January 15, 1967.

Kraus, Jack. *Quote, The Weekly Digest*. February 5, 1967.

Kraus, Jack. *Quote, The Weekly Digest*. April 2, 1967.

Kraus, Jack. *Quote, The Weekly Digest*. August 4, 1968.

Kraus, Karl. *Half-Truths & One-and-a-Half Truths*. Engenda Press, Montreal. 1976.

La Bruyère, Jean. *The Characters*. Translated by Henri Van Laun. Oxford University Press, London. 1963.

La Rochefoucauld, François de. *Maxims of La Rochefoucauld*. The Hawthorne Press. 1931.

Lamb, Charles. *Essays of Elia*. G.P. Putnam's Sons, New York. 1884.

Lamb, Charles. *The Complete Works and Letters of Charles Lamb*. The Modern Library, New York. 1935.

Lamont, Thomas W. *My Boyhood in a Parsonage*. Harper & Brothers Publishers, New York. 1946.

Langer, Susan. *Philosophy in a New Key*. New American Library, New York. 1951.

Lash, Joseph P. *Eleanor: The Years Alone*. New American Library, New York. 1973.

Lawrence, Robert Means. *Primitive Psycho-Therapy and Quackery*. Houghton Mifflin Company, Boston. 1910.

Legman, G. *The New Limerick*. Crown Publishers, New York. 1977.

Legrain, G. *Répertoire Généalogique Et Onomastique Du Musée Du Caire*. Société Anonyme des Arts Graphiques, Genève. 1908.

Lehrs, Ernst. *Man or Matter*. Faber and Faber, Ltd., London. 1951.

Lerner, Max. *The Unfinished Country*. Simon and Schuster, New York. 1959.

Lewis, C.S. *The Screwtape Letters*. The Macmillan Company, New York. 1948.

Lewis, Lucille. 'This I Believe' in *Nursing Outlook*. Volume 15, Number 5. May 1968.

Lewis, Sinclair. *Arrowsmith*. The Modern Library, New York. 1925.

Lippmann, Walter. *Liberty and the News*. Harcourt, Brace and Howe, New York. 1920.

Livermore, Mary A. *What Shall We Do With Our Daughters?* Lee and Shepard Publishers, Boston. 1883.

Lomax, John Avery. *American Ballads and Folk Songs*. Macmillan, New York. 1934.

Longfellow, Henry Wadsworth. *The Song of Hiawatha*. Ticknor and Fields, Boston. 1856.

Longfellow, Henry Wadsworth. *The Works of Henry Wadsworth Longfellow*. Volume V. Fireside Edition, Boston. 1910.

Longford, Elizabeth. *Queen Victoria: Born to Succeed*. Harper & Row Publishers, New York. 1964.

Luminet, Jean-Pierre. *Black Holes*. University Press, Cambridge. 1987.

Lyly, John. *Euphues*. Southgate, London. 1868.

Mansfield, Katherine. *The Journal of Katherine Mansfield*. Alfred A. Knopf, New York. 1927.

Marquis, Don. *the lives and times of archy & mehitabel*. Doubleday Doran & Co., Inc., Garden City. 1933.

Martial. *Epigrams*. Volume I. Translated by Walter C.A. Ker. Harvard University Press, Cambridge. 1947.

Massengill, Samuel Evans. *A Sketch of Medicine and Pharmacy*. The S.E. Massengill Company, Bristol. 1943.

Massinger, Philip. *A New Way to Pay Old Debts*. The Athlone Press, University of London. 1956.

Massinger, Philip. *The Bondman*. Princeton University Press, Princeton. 1932.

Massingill, Philip. *The Plays of Philip Massinger*. Volume I. AMS Press, New York. 1966.

Mather, Cotton. *The Angel of Bethesda*. Barre Publishers, Barre. 1972.

Mather, Increase. *Remarkable Providences*. Reeves and Turner, London. 1890.

Matthews, Marian. 'Interns' in *American Journal of Nursing*. November 1968.

Maturin, Charles R. *Bertram*. Printed for John Murray, London. 1816.

Mayo, Charles H. 'Problems of Infection' in *Minnesota Medicine*. Volume 1. 1918.

Mayo, Charles H. 'Carcinoma of the Right Segment of the Colon' in *Annals of Surgery*. Volume 83. March 1926.

Mayo, Charles H. 'The Value of Broadmindedness' in *Medical Life*. Volume 34. April 1927.

Mayo, Charles H. 'La funcion del higado en relacion con la cirugia' in *Annals de Circulation*. Volume 2. April 1930.

Mayo, Charles H. 'When Does Disease Begin? Can This Be Determined by Health Examination?' in *Minnesota Medicine*. Volume 15. January 1932.

Mayo, Charles H. *Modern Hospital*. Volume 51. September 1938.

Mayo, William J. 'The Cancer Problem' in *Journal-Lancet*. Volume 35, July 1, 1915.

Mayo, William J. 'In the Time of Henry Jacob Bigelow' in *Journal of the American Medical Association*. Volume 77. August 20, 1921.

Mayo, William J. 'The Influence of Ignorance and Neglect on the Incidence and Mortality of Cancer' in *Journal of the Indiana Medical Association*. Volume 17. 1924.

Mayo, William J. 'The Teaching Hospital of the University of Michigan' in *Journal of the Michigan Medical Society*. Volume 25. January 1926.

Mayo, William J. 'Looking Backward and Forward in Medical Education' in *Journal of the Iowa Medical Society*. Volume 19. February 1929.

Mayo, William J. 'Master Surgeons of America; Frederic S. Dennis' in *Surgery, Gynecology and Obstetrics*. Volume 67. October 1938.

McCullers, Carson. *Clock Without Hands*. Houghton Mifflin Company, Boston. 1961.

McDonough, Mary Lou. *Poet Physicians*. Charles C. Thomas, Springfield. 1945.

McElwee, Tom. *Quote, The Weekly Digest*. June 2, 1968.

McLaughlin, Mignon. *The Neurotic's Notebook*. The Bobbs-Merrill Company, Inc., Indianapolis. 1963.

Melton, John. *Astrologaster*. The Augustan Reprint Society, Los Angeles. 1975.

Melville, Helen and Melville, Lewis. *An Anthology of Humorous Verse*. Dodd, Mead & Company, New York. No date.

Melville, Herman. *Moby Dick* in *Great Books of the Western World*. Volume 32. Encyclopaedia Britannica, Inc., Chicago. 1952.

Mencken, H.L. *A New Dictionary of Quotations on Historical Principles*. Alfred A. Knopf, New York. 1942.

Mencken, H.L. *Prejudices: Third Series*. A.A. Knopf, New York. 1922.

Mencken, H.L. 'What is Going on in the World' in *The American Mercury*. Volume XXX, Number 119. November 1933.

Metcalf, Elizabeth. 'Picturesque Speech and Patter' in *Reader's Digest*. September 1946.

Meyer, Adolf. 'The "Complaint" As the Center of Genetic–Dynamic and Nosological Teaching In Psychiatry' in *New England Journal of Medicine*. August 23, 1928.

Middleton, Thomas. *Women Beware Women*. University of California Press, Berkeley. 1969.

Milton, John. *Paradise Lost* in *Great Books of the Western World*. Volume 32. Encyclopaedia Britannica, Inc., Chicago. 1952.

Milton, John. *Samson Agonistes*. Oxford University Press, Oxford. 1957.

Milton, John. *The Reason of Church-Government*. London. 1641.

Mitchell, Margaret. *Gone With the Wind*. The Macmillan Company, New York. 1936.

Mitford, Nancy. *Noblesse Oblige*. Hamish Hamilton, London. 1956.

Montaigne, Michel de. *Essays* in *Great Books of the Western World*. Volume 25. Encyclopaedia Britannica, Inc., Chicago. 1952.

Moore, Thomas. *The Poetical Works of Thomas Moore*. William Collins, Sons & Co., Limited, London. No date.

Morley, Christopher. *Human Being*. Doubleday, Doran & Company, Inc., Garden City. 1932.

Morris, Joseph F. *Quote, The Weekly Digest*. July 21, 1968.

Morris, Robert Tuttle. *Doctors versus Folks*. Doubleday, Page and Company, Garden City. 1915.

Muller, Herbert J. *Science and Criticism*. Yale University Press, New Haven. 1943.

Nash, Ogden. 'Have You A Pash for Ogden Nash' in *The Reader's Digest*. July 1952.

Newman, Sir George. 'Preventive Medicine for the Medical Student' in *The Lancet*. Volume 221. November 21, 1931.

Nightingale, Florence. *Notes on Nursing*. Harrison, London. 1946.

Nolan, James Joseph. 'On Renewed Maternal Mortality Reports' in *The New England Journal of Medicine*. Volume 286, Number 17. April 27, 1972.

Nye, Bill. *Remarks*. G.P. Brown Publishing Co., Chicago. 1888.

O'Neill, Eugene. *Long Day's Journey into Night*. Yale University Press, New Haven. 1956.

Ogden, C.K. and Richards, I.A. *The Meaning of Meaning*. Harcourt, Brace and Company, London. 1952.

Ogilvie, Sir Heneage. 'A Surgeon's Life' in *The Lancet*. Volume 255. July 3, 1948.

Ogutsch, Edith. *Quote, The Weekly Digest*. May 7, 1967.

Osler, Sir William. *Aequanimitas*. The Blakiston Company, Philadelphia. 1942.

Osler, Sir William. *Evolution of Modern Medicine*. Yale University Press, New Haven. 1921.

Osler, Sir William. *Teacher and Student*. John Murphy & Co., Baltimore. 1892.

Osler, Sir William. *The Master-Word in Medicine*. John Murphy Company, Baltimore. 1903.

Ott, Susan. 'A Pulmonologist's Valentine' in *The New England Journal of Medicine*. Volume 304, Number 12. 1981.

Ovid. *Ex Ponto*. Harvard University Press, Cambridge. 1959.

Ovid. *Metamorphoses*. Translated by Rolfe Humphries. Indiana University Press, Bloomington. 1958.

Ovid. *Tristia*. Harvard University Press, Cambridge. 1959.

Oyle, Irving. *The New American Medical Show*. Unity Press, Santa Cruz. 1925.

Paget, Stephen. *Confessio Medici*. The Macmillan Company, New York. 1931.

Pascal, Blaise. *Penseés* in *Great Books of the Western World*. Volume 33. Encyclopaedia Britannica, Inc., Chicago. 1952.

Pasveer, B. 'Knowledge of Shadows: The Introduction of X-ray Images in Medicine' in *Sociology of Health and Illness*. Volume 11. 1989.

Paulos, John Allen. *A Mathematician Reads the Newspaper*. Basic Books, New York. 1995.

Peabody, Francis Weld. *The Care of the Patient*. Harvard University Press, Cambridge. 1928.

Pearl, Cyril. *The Best of Lennie Lower*. Lansdowne Press, Melbourne. 1963.

Pekkanen, John. *M.D.—Doctors Talk About Themselves*. Delacorte Press, New York. 1988.

Percival, Thomas. *Medical Ethics*. S. Russell, Manchester. 1803.

Perelman, S.J. *Crazy Like a Fox*. Random House, New York. 1944.

Persius. *The Satires of Persius*. Indiana University Press, Bloomington. 1961.

Petronius. *The Satyricon*. New American Library, New York. 1959.

Pickering, James Sayre. *The Stars are Yours*. Macmillan, New York. 1953.

Pitkin, Walter B. *The Twilight of the American Mind*. Simon and Schuster, New York. 1928.

Planck, Max. *Scientific Autobiography*. Philosophical Library, New York. 1949.

Plath, Sylvia. *The Bell Jar*. Harper & Row, Publishers, New York. 1971.

Plato. *Phaedrus* in *Great Books of the Western World*. Volume 7. Encyclopaedia Britannica, Inc., Chicago. 1952.

Plato. *Symposium* in *Great Books of the Western World*. Volume 7. Encyclopaedia Britannica, Inc., Chicago. 1952.

Plato. *The Republic* in *Great Books of the Western World*. Volume 7. Encyclopaedia Britannica, Inc., Chicago. 1952.

Plutarch. *The Lives of the Noble Grecians and Romans* in *Great Books of the Western World*. Volume 14. Encyclopaedia Britannica, Inc., Chicago. 1952.

Poe, Edgar Allan. *The Raven*. Columbia University Press, New York. 1942.

Pope, Alexander. *An Essay on Criticism*. Sidhartha Publications, Karnal. 1988.

Pope, Alexander. *The Poems of Alexander Pope*. Volume III. Methuen & Co., Ltd., London. 1953.

Porter, Roy. *The Greatest Benefit to Mankind*. HarperCollins Publishers, Hammersmith. 1997.

Potter, Stephen. *One-Upmanship*. Henry Holt and Company, New York. 1952.

Pound, Louise. *American Ballads and Songs*. C. Scribner's Sons, New York. 1972.

Prentice, George Denison. *Prenticeana*. Derby & Jackson, New York. 1860.

Preston, Margaret Junkin. *Old Songs and New*. J.B. Lippincott & Co., Philadelphia. 1870.

Prior, Matthew. *The Poetical Works of Matthew Prior*. Bell and Daldy, London. 1800.

Prochnow, Herbert V. and Prochnow, Herbert V., Jr. *A Treasury of Humorous Quotations*. Harper and Row, Publishers, New York. 1969.

Propp, Fred Jr. *Quote, The Weekly Digest*. July 23, 1967.

Proust, Marcel. *The Guermantes Way*. Translated by C.K. Scott Moncrieff. The Modern Library, New York. 1925.

Quarles, Francis. *Hieroglyphikes*. Scholar Press, Menston. 1969.

Rabelais, François. *Gargantua* in *Great Books of the Western World*. Volume 24. Encyclopaedia Britannica, Inc., Chicago. 1952.

Rabelais, François. *Pantagruel* in *Great Books of the Western World*. Volume 24. Encyclopaedia Britannica, Inc., Chicago. 1952.

Ramón y Cajal, Santiago. *Recollections of My Life*. Translated by E. Horn Craigie. The M.I.T. Press, Cambridge. 1960.

Ransom, John Crowe. *The Poetry of John Crowe Ransom*. Rutgers University Press, New Brunswick. 1972.

Ray, John. *A Complete Collection of English Proverbs*. Printed for George Cowie and Co., London. 1813.

Reagan, Ronald. 'Anderson and Reagan Disagree on Many Issues, but Treat Carter Lightly' in *New York Times*, Section B, page 5, Column 2, September 22, 1980.

Redford, Sophie, E. 'To the D.D.S.' in *Cartoon Magazine*. Volume 18, Number 6. December 1920.

Reichen, Charles-Albert. *A History of Astronomy*. Hawthorn Books Inc., New York. 1963.

Reid, Thomas. *The Works of Thomas Reid, D.D.* Maclachlan and Stewart, Edinburgh. 1880.

Renya, Ruth. *The Philosophy of Matter in the Atomic Era*. Asia Publishing House, Bombay. 1962.

Richardson, Samuel. *The History of Sir Charles Grandison*. Oxford University Press, London. 1972.

Richardson, Samuel. *The Works of Samuel Richardson*. Volume VII. *The History of Clarissa Harlowe*. Volume IV. Henry Sothern & Co., London. 1883.

Richler, Mordecai. *Son of a Smaller Hero*. McClelland and Stewart Limited, Toronto. 1955.

Rogers, Will. *The Autobiography of Will Rogers*. Houghton Mifflin Company, Boston. 1949.

Romains, Jules. *Knock*. Translated by James B. Gidney. Barron's Educational Series, Inc., Great Neck. 1959.

Romanoff, Alexis Lawrence. *Encyclopedia of Thoughts*. Ithaca Heritage Books, Ithaca. 1975.

Rous, Francis. *Thule*. Burt Franklin, New York. 1967.

Roy, Gabrielle. *The Cashier*. Harcourt, Brace and Company, New York. 1955.

Ruskin, John. *The Crown of Wild Olives & the Cestus of Aglaia*. J.M. Dent & Sons Ltd., London. 1915.

Russell, Bertrand. *The ABC of Atoms*. Kegan Paul, Trench, Trubner & Co., Ltd., London. 1927.

Russell, L.K. 'Line on an X-ray: Portrait of a Lady' in *Life*. March 12, 1896.

Sacks, Oliver. 'Listening to the Lost' in *Newsweek*. August 20, 1984.

Sacks, Oliver. *Awakenings*. E.P. Dutton, Inc., New York. 1983.

Saint Augustine. *Confessions* in *Great Books of the Western World*. Volume 18. Encyclopaedia Britannica, Inc., Chicago. 1952.

Saint Augustine. *The City of God* in *Great Books of the Western World*. Volume 18. Encyclopaedia Britannica, Inc., Chicago. 1952.

Sanger, Margaret. *My Fight for Birth Control*. Farrar & Rinehart, Inc., New York. 1931.

Schmitz, Jacqueline T. 'Point of View' in *American Journal of Nursing*. November 1968.

Scott, Sir Walter. *The Surgeon's Daughter*. George Routledge & Sons, London. 1879.

Scott, Sir Walter. *The Talisman*. Thomas Nelson & Sons Ltd., London. No date.

Selden, John. *Table Talk of John Selden*. Cassell & Company, Ltd., London. 1887.

Selye, Hans. *From Dream to Discovery*. McGraw-Hill Book Company, New York. 1964.

Selzer, Richard. *Mortal Lessons*. Simon and Schuster, New York. 1976.

Seneca. *Clemency*. Printed by Thomas Harper, London. 1653.

Seneca. *Ad Lucilium Epistulae Morales*. William Heinemann, London. 1917.

Shadwell, Thomas. *The Complete Works of Thomas Shadwell*. The Fortune Press, London. 1927.

Shakespeare, William. *A Midsummer-Night's Dream* in *Great Books of the Western World*. Volume 26. Encyclopaedia Britannica, Inc., Chicago. 1952.

Shakespeare, William. *As You Like It* in *Great Books of the Western World*. Volume 26. Encyclopaedia Britannica, Inc., Chicago. 1952.

Shakespeare, William. *Coriolanus* in *Great Books of the Western World*. Volume 27. Encyclopaedia Britannica, Inc., Chicago. 1952.

Shakespeare, William. *Cymbeline* in *Great Books of the Western World*. Volume 27. Encyclopaedia Britannica, Inc., Chicago. 1952.

Shakespeare, William. *Hamlet* in *Great Books of the Western World*. Volume 27. Encyclopaedia Britannica, Inc., Chicago. 1952.

Shakespeare, William. *Julius Caesar* in *Great Books of the Western World*. Volume 26. Encyclopaedia Britannica, Inc., Chicago. 1952.

Shakespeare, William. *King Henry the Eighth* in *Great Books of the Western World*. Volume 27. Encyclopaedia Britannica, Inc., Chicago. 1952.

Shakespeare, William. *The First Part of King Henry the Fourth* in *Great Books of the Western World*. Volume 26. Encyclopaedia Britannica, Inc., Chicago. 1952.

Shakespeare, William. *The First Part of King Henry the Sixth* in *Great Books of the Western World*. Volume 26. Encyclopaedia Britannica, Inc., Chicago. 1952.

Shakespeare, William. *The Life and Death of King John* in *Great Books of the Western World*. Volume 26. Encyclopaedia Britannica, Inc., Chicago. 1952.

Shakespeare, William. *King Lear* in *Great Books of the Western World*. Volume 27. Encyclopaedia Britannica, Inc., Chicago. 1952.

Shakespeare, William. *Macbeth* in *Great Books of the Western World*. Volume 27. Encyclopaedia Britannica, Inc., Chicago. 1952.

Shakespeare, William. *Much Ado About Nothing* in *Great Books of the Western World*. Volume 26. Encyclopaedia Britannica, Inc., Chicago. 1952.

Shakespeare, William. *Othello* in *Great Books of the Western World*. Volume 27. Encyclopaedia Britannica, Inc., Chicago. 1952.

Shakespeare, William. *The Rape of Lucrece*. Winchester Publications Limited, London. 1948.

Shakespeare, William. *Romeo and Juliet* in *Great Books of the Western World*. Volume 26. Encyclopaedia Britannica, Inc., Chicago. 1952.

Shakespeare, William. *The Tempest* in *Great Books of the Western World*. Volume 27. Encyclopaedia Britannica, Inc., Chicago. 1952.

Shakespeare, William. *Timon of Athens* in *Great Books of the Western World*. Volume 27. Encyclopaedia Britannica, Inc., Chicago. 1952.

Shakespeare, William. *Troilus and Cressida* in *Great Books of the Western World*. Volume 27. Encyclopaedia Britannica, Inc., Chicago. 1952.

Shakespeare, William. *Twelfth Night* in *Great Books of the Western World*. Volume 27. Encyclopaedia Britannica, Inc., Chicago. 1952.

Shakespeare, William. *The Two Gentlemen of Verona* in *Great Books of the Western World*. Volume 26. Encyclopaedia Britannica, Inc., Chicago. 1952.

Shapp, Paul. 'The Fastest Man on Earth' in *Time*. Volume LXVI, Number 11. September 12, 1955.

Sharp, Cecil. *English Folk Songs from the Southern Appalachians*. Oxford University Press, London. 1960.

Sharpe, Tom. *Porterhouse Blue*. Secker & Warburg, London. 1974.

Shaw, George Bernard. *Back to Methuselah*. Brentano's, New York. 1927.

Shaw, George Bernard. *Man and Superman*. The Heritage Press, New York. 1962.

Shaw, George Bernard. *Misalliance*. Brentano's, New York. 1930.

Shaw, George Bernard. *The Doctor's Dilemma*. Brentano's Publishers, New York. 1911.

Shaw, George Bernard. *The Revolutionist's Handbook & Pocket Companion*. United States. 1962?

Sheridan, Richard Brinsley. *St. Patrick's Day*. Printed for the Bookseller, Dublin. 1788.

Siegel, Eli. *Damned Welcome*. Definition Press, New York. 1972.

Sigerist, Herny E. 'Bedside Manners in the Middle Ages: The Treatise De Cautelis Medicorum Attributed to Arnald of Villanova' in *Quarterly Bulletin of Northwestern University Medical School*. Volume 20. 1946.

Simmons, Charles. *Laconic Manual and Brief Remarker*. Robert Dick, Toronto. 1853.

Skelton, John. *The Complete English Poems*. Yale University Press, New Haven. 1983.

Smith, H. Allen. *Quote, The Weekly Digest*. February 19, 1967.

Smith, Homer W. *From Fish to Philosopher*. Little, Brown and Company, Boston. 1953.

Smith, Theobald. 'The Influence of Research in Bringing into Closer Relationship the Practice of Medicine and Public Health Activities' in *American Journal of the Medical Sciences*. December 1929.

Smollett, Tobias. *Sir Launcelot Greaves*. Oxford University Press, London. 1973.

Södergran, Edith. *We Women*. Translated by Samuel Charters. Oyez, Berkeley. 1977.

Sontag, Susan. *Illness as Metaphor*. Farrar, Straus and Giroux, New York. 1978.

Sophocles. *Oedipus at Colonus*. Harcourt, Brace and Company, New York. 1941.

Southerne, Thomas. *The Loyal Brother*. Printed for William Cademon, London. 1692.

Spence, Sir James. *The Purpose and Practice of Medicine*. Oxford University Press, London. 1960.

Staff. *National Lampoon Tenth Anniversary Anthology 1970–1980*. National Lampoon, Inc., New York. 1979.

Stanton, Elizabeth Cady. *Eighty Years and More*. T. Fisher Unwin, London. 1898.

Steinbeck, John. *East of Eden*. Viking Press, New York. 1952.

Stern, B.J. *Society and Medical Progress*. Princeton University Press, Princeton. 1941.

Sterne, Laurence. *A Sentimental Journey through France and Italy*. Oxford University Press, London. 1928.

Sterne, Laurence. *Tristram Shandy*. Shakespeare Head Press, Stratford-Upon-Avon. 1926.

Stevenson, Robert Louis. *Treasure Island*. Ginn and Company, Boston. 1911.

Stevenson, Robert Louis. *Underwoods*. Chatto and Windus, London. 1887.

Stewart, Michael M. 'Help?' in *The New England Journal of Medicine*. Volume 285, Number 24. 1971.

Stewart, Michael M. 'Manpower Planning (By Degrees)' in *The New England Journal of Medicine*. Volume 287, Number 13. September 28, 1972.

Stumpf, LeNore. 'Needed: Remote Control' in *American Journal of Nursing*. April 1969.

Sutton, DeWitt, Jr. 'Gout & Metabolism' in *American Scientist*. Volume 198, Number 6. June 1958.

Swift, Jonathan. *Polite Conversation* in *The Prose Works of Jonathan Swift*. Volume the Fourth. Shakespeare Head Press, Oxford. 1957.

Swift, Jonathan. *Satires and Personal Writings*. Oxford University Press, London. 1944.

Syrus, Publilius. *Sententiæ*. Lipsiae, in Aedibus B.G. Teubneri. 1880.

Szasz, Thomas. *The Second Sin*. Anchor Press, Garden City. 1973.

Taylor, Jeremy. *Holy Living and Holy Dying*. Volume II. Clarendon Press, Oxford. 1989.

Taylor, John G. *The New Physics*. Basic Books, Inc., New York. 1972.

Teilhard de Chardin, Pierre. *The Phenomenon of Man*. Harper & Row Publisher, New York. 1959.

Teller, Joseph D. 'Floaters and Sinkers' in *The New England Journal of Medicine*. Volume 287, Number 1. 1972.

Tertullian. *Tertullian's Apology*. Printed by Tho. Harper, London. 1655.

Thackery, William Makepeace. *The History of Pendennis*. Charles Scribner's Sons, New York. 1917.

Thomas, Gwyn. *The Keep*. Elek Books, London. 1962.

Thompson, Francis. *Complete Poetical Work of Francis Thompson*. Boni and Liveright, Inc., New York. 1913.

Thomson, James. *The Castle of Indolence and Other Poems*. University of Kansas Press, Lawrence. 1961.

Thoreau, Henry David. *A Week on the Concord and Merrimac Rivers*. Walter Scott Limited, London. 1889.

Thurber, James. *The Thurber Carnival*. Harper & Brothers, New York. 1931.

Tolstoy, Leo. *War and Peace* in *Great Books of the Western World*. Volume 21. Encyclopaedia Britannica, Inc., Chicago. 1952.

Trilling, Lionel. *The Liberal Imagination*. Doubleday Anchor Books, Garden City. 1950.

Trotter, Wilfred. *Collected Papers*. Oxford University Press, London. 1941.

Twain, Mark. *A Tramp Abroad*. Harper & Brothers Publishers, New York. 1921.

Twain, Mark. *Europe and Elsewhere*. Harper & Brothers, Publishers, New York. 1923.

Twain, Mark. *Following the Equator*. The American Publishing Company, Hartford. 1898.

Twain, Mark. *Letters From The Earth*. Harper and Row, Publishers, New York. 1962.

Twain, Mark. *Mark Twain's Travels with Mr. Brown*. Alfred A. Knopf, New York. 1946.

Twain, Mark. *The Guilded Age*. Nelson Doubleday, Inc., Garden City. No date.

Twain, Mark. *The Tragedy of Pudd'nhead Wilson*. Oxford University Press, New York. 1996.

Unamuno, Miguel de. *Essays and Soliloquies*. Translated by J.E. Crawford Flitch. George G. Harrap & Company, Ltd., London. 1925.

Untermeyer, Louis. *Lots of Limericks*. Doubleday & Company, Inc., Garden City. 1961.

Viereck, George Sylvester and Eldridge, Paul. *My First Two Thousand Years*. Sheridan House, Inc., New York. 1963.

Virchow, Rudolf. *Disease, Life, and Man*. Translated by Lelland J. Rather. Stanford University Press, Stanford. 1958.

Voltaire. *Philosophical Dictionary*. Volume VII. E.R. SuMont, Paris. 1901.

Voltaire. *Philosophical Dictionary*. Volume VIII. E.R. SuMont, Paris. 1901.

von Ebner-Eschenbach, Marie. *Aphorisms*. Translated and Introduced by David Scrase and Wolfgang Mieder. Ariadne Press, Riverside. 1994.

Walis, Claudia. 'Today's Dentistry: A New Drill' in *Time*. September 9, 1985.

Wallace, Marilyn. *Sisters in Crime*. Volume I. Berkley Books, New York. 1989.

Walton, Izaak. *The Compleat Angler*. Nathaniel Cooke, Strand. 1854.

Ward, Artemus. *The Complete Works of Artemus Ward*. G.W. Dillingham Co., Publisher, New York. 1893.

Warren, Roz. *Glibquips*. The Crossing Press, Freedom. 1994.

Watson, William. *The Poems of William Watson*. Macmillan and Co., New York. 1893.

Watts, Alan. *The Wisdom of Insecurity*. Pantheon Books, Inc., New York. 1951.

Waugh, Evelyn. *Noblesse Oblige*. Greenwood Press, Publishers, West Port. 1956.

Waugh, Evelyn. *Vile Bodies*. Jonathan Cape Harrison Smith, New York. 1930.

Weaver, Jefferson Hane. *The World of Physics*. Volume II. Simon and Schuster, New York. 1987.

Webster, John. *The Duchess of Malfi*. Washington Square Press, Inc., New York. 1959.

Webster, John. *The White Devil*. University of Nebraska Press, Lincoln. 1969.

Weingarten, Violet. *Intimations of Mortality*. Alfred A. Knopf, New York. 1978.

Wells, H.G. *Bealby*. The Macmillan Company, New York. 1913.

Wells, H.G. *Experiment in Autobiography*. The Macmillan Company, New York. 1934.

Wells, H.G. *Meanwhile*. George H. Doran Company, New York. 1927.

Wells, H.G. *The Undying Fire*. The Macmillan Company, New York. 1919.

Welsh, Joan I. *Quote, The Weekly Digest*. July 21, 1968.

Welty, Eudora. *Selected Stories of Eudora Welty*. The Modern Library, New York. 1966.

Weyl, Hermann. *The Theory of Groups and Quantum Mechanics*. Methuen & Co., Ltd., London. 1931.

Wheeler, Hugh. *A Little Night Music*. Dodd, Mead & Company, New York. 1973.

Whitehead, Alfred North. *An Introduction to Mathematics*. Oxford University Press, London. 1948.

Wilde, Oscar. *The Importance of Being Earnest*. Appeal to Reason, Girard. 1921.

Wilde, Oscar. *The Works of Oscar Wilde*. Collins, London. 1946.

Wilder, Thornton. *The Bridge of San Luis Rey*. Albert & Charles Boni, New York. 1928.

Willstätter, Richard. *From My Life*. W.A. Benjamin, Inc., New York. 1965.

Wilson, Edward O. *Biophilia*. Harvard University Press, Cambridge. 1984.

Wodehouse, P.G. *The Inimitable Jeeves*. Penguin Books Ltd., Middlesex. 1953.

Wolfe, Humbert. *Cursory Rhymes*. Ernest Benn Limited, London. 1927.

Wolfe, Thomas. *Look Homeward, Angel*. The Modern Library, New York. 1929.

Woolf, Virginia. *The Moment*. Harcourt, Brace and Company, New York. 1948.

Wordsworth, William. *The Complete Poetical Works of William Wordsworth*. Macmillan and Co., London. 1891.

Wright, Frank Lloyd. 'Frank Lloyd Wright and His Art' in *New York Times Magazine*. October 4, 1953.

Wycherley, William. *The Country Wife*. Random House, New York. 1936.

Young, Arthur. *The Adventures of Emmera*. Garland Publishing, Inc., New York. 1974.

Young, Arthur. *Travels in France*. At the University Press, Cambridge. 1950.

Young, Edward. *Love of Fame*. Printed for J. Tonson in the Strand, London. 1728.

Young, Edward. *Night Thoughts*. Dover Publications, Inc., New York. 1975.

Yourcenar, Marguerite. *Memoirs of Hadrian*. Farrar, Straus and Young, New York. 1954.

PERMISSIONS

SUBJECT BY AUTHOR INDEX

anatomists
Richardson, Samuel
  And I believe that anatomists
    allow..., 9
Twain, Mark
  Anatomists see no beautiful
    women in all their lives..., 9
anatomy
Bacon, Francis
  In the inquiry which is made by
    anatomy..., 10
Burton, Robert
  ...make them as so many
    anatomies..., 10
Dagi, Teodoro Forcht
  ...And say "to pass anatomy", 10
Dickinson, Emily
  A science—so the Savants say,
    10
Fernel, Jean
  Anatomy is to physiology..., 11
Halle, John
  But chieflye the anatomye..., 11
Holmes, Oliver Wendell
  ...that method of study to which
    is given the name of General
    Anatomy..., 11
Muller, Herbert J.
  To say that a man is made up of
    certain chemical elements...,
    11
Nye, Bill
  The word anatomy is derived
    from two Greek spatters...,
    11
  Human anatomy is either..., 11
Osler, Sir William
  Anatomy may be likened to a
    harvest field..., 12
Reid, Thomas
  ...without any previous
    knowledge of anatomy..., 12
Shapp, Paul

  The human body comes in
    only two shapes and three
    colors..., 12
anesthesia
du Bartas, Guillaume de Saluste
  Even as a surgeon, minding off
    to cut/Some cureless limb...,
    13
Genesis 2:21
  And the Lord God caused a
    deep sleep to fall upon
    Adam..., 13
Helmuth, William Tod
  Adam profoundly slept with
    anesthesia..., 13
Kraus, Karl
  Anesthesia: wounds without
    pain, 14
Massinger, Philip
  A sleep potion, that will hold
    her long..., 14
anesthetics
Holmes, Oliver Wendell
  ...three natural anaesthetics..., 14
anesthetist
Trotter, Wilfred
  Mr. Anaesthetist, if the patient
    can keep awake..., 15
anesthetist's cone
Cvikota, Raymond J.
  Anesthetist's cone: Ether
    bonnet, 15
anesthetized
Armour, Richard
  .../Not fighting
    back—anesthetized, 13
antibiotics
Unknown
  What to give a man who has
    everything..., 16
apothecary
Bierce, Ambrose
  Disease for the apothecary's
    health..., 17

AUTHOR BY SUBJECT INDEX

-A-

Abse, Dannie (1923–)
Poet
  tumor, 370
Ace, Goodman
  hypochondriac, 171
  medical advice, 197
Adams, Cedric
  ills, 175
Adams, Henry Brooks (1838–1918)
Historian
  experience, 137
Advertisement
  cure, 57
Albricht, Fuller
  doctor, 99
Alexander the Great (356–323 BC)
King of Macedonia
  physician, 252
Alexander, Franz (1891–1964)
Physician
  cure, 57
Alison, Richard (1588–1606)
Writer
  teeth, 352
Allbutt, Sir Thomas Clifford
  (1836–1925)
English physician
  science, 312
Allen, Woody (1935–)
US Film maker
  contraception, 53

Allman, David
  death, 63
  physician, 252
Amiel, Henri-Frédéric (1821–1881)
Swiss philosopher
  curing, 57
  doctor, 99
  error, 134
Anderson, Peggy
  nurses, 224
Aristotle (384–322 BC)
Greek philosopher
  physician, 252
Armour, Richard (1906–)
  adhesive, 5
  anesthetized, 13
  dermatologist, 77
  doctor, 99
  druggist, 127
  enema, 131
  patient, 241
  tongue depressor, 358
  urinalysis, 371
Armstrong, John (1709–1779)
  blood, 26
  brain, 32
  medical, 197
Arnauld, Antoine (1612–1694)
French theologist
  common sense, 49
  science, 312

Arnold, Matthew (1822–1888)
English Victorian poet
  doctor, 100
Arnoldus
  diet, 85
Askey, Vincent
  hypochondriacs, 171
Aurelius, Marcus [Antoninus]
  physician, 252
Aylett, Robert (1583–1685)
  surgeon, 339

-B-

Bacon, Francis (1561–1626)
English statesman
  anatomy, 10
  diet, 85
  diseases, 89
  doctor, 100
  medicine, 203
  physician, 252
Bahya, ben Joseph ibn Paauda
    (Second half 11th century)
Judge of a rabbinical court
  patient, 241
Bailey, Percival (1740–1804)
  physician, 253
Baldwin, Joseph G.
  physician, 253
Balzac, Honoré de (1799–1850)
  doctor, 100
Barach, Alvin
  disease, 89
Baring, Maurice
  disease, 89
Barnard, Christiaan N. (1922–)
South African surgeon
  heart, 164
Barnes, Djuna (1892–1982)
  arteries, 20
  heart, 164
  nurses, 224
  sickness, 320
Barss, Peter
  injury, 179

Baruch, Bernard (1870–1965)
Financier
  cure, 57
Bass, Murray H.
  physician, 253
Bassler, Thomas J.
  death, 63
Bates, Marston (1960–)
American zoologist/science journalist
  research, 306
Bates, Rhonda
  PMS, 286
Battles, William Snowden
  hypodermic needle, 174
Baum, Harold
  nutrition, 231
Baxter, Richard (1615–1691)
English theologian
  tooth, 352
Bayliss, William Maddock
  physiologist, 281
Beacock, Cal
  flu, 149
Beard, George M.
  error, 134
Beaumont, Francis (1584–1616)
British dramatist
  cure, 58
Beckett, Samuel (1906–1989)
Irish novelist, dramatist, poet
  hospital, 168
  nurses, 224
Bell, H.T.M. (1856–1930)
  dentist, 70
Belloc, Hilaire (1870–1953)
French-born poet/essayist
  disease, 89
Benchley, Robert (1889–1945)
American drama critic
  cold, 47
Benjamin, Arthur
  dentist, 70
Berkenhout, John
  medicine, 203
Bernard, Claude (1813–1878)

fee, 145
quacks, 296
sickness, 321
teeth, 354
Hooker, Worthington
  facts, 143
Howard, Sidney (1891–1931)
US playwright
  doctor, 110
Howell, James (1594–1666)
Anglo-Welsh writer
  pain, 238
Howells, William Dean
    (1837–1920)
US novelist
  physician, 263
Hubbard, Elbert (1859–1915)
US author/editor
  back, 21
  doctor, 263
  surgery, 346
Hubbard, Kin
  doctor, 110
  general practitioners, 151
  patient, 243
Hufeland, Christoph Wilhelm
  physician, 263
Hulme, Keri
  medicine, 209
Hunter, William (1718–1783)
British obstetrician
  stomach, 336
Husserl, E. (1859–1938)
German philosopher
  fact, 143
Hutchison, Sir Robert
  diagnosis, 80
  diagnosticians, 83
  teaching, 350
Huxley, Aldous (1894–1963)
English novelist
  microbes, 217
Huxley, Julian (1887–1975)
English biologist
  brain, 32

Huxley, Thomas Henry
    (1825–1895)
English naturalist
  medicine, 209
  physiology, 282
  vaccination, 372

-I-
Inge, W.R. (1860–1954)
British Dean of St. Paul's Cathedral
  cholera, 43
Isaiah
  sick, 317

-J-
Jackson, James
  drugs, 128
  physician, 263
James, Alice (1848–1892)
  doctor, 110
James, Henry (1843–1916)
American novelist/critic
  disease, 92
  toothache, 360
James, William (1842–1910)
American philosopher
  common sense, 50
Janssen, Johannes (1829–1891)
German historian
  chemistry, 40
Jauncey, G.E.M.
  x-rays, 374
Jefferson, Thomas (1743–1826)
US president
  error, 134
Jekyll, Joseph
  physician, 264
Jenner, Edward (1749–1823)
English surgeon
  quacks, 297
Jeremiah
  physician, 264
Jerome, Jerome K. (1859–1927)
English writer
  diseases, 92

La Rochefoucauld, François de (1613–1680)
French moralist
  health, 160
Lamb, Charles (1775–1834)
English essayist
  convalescence, 55
  sick-bed, 319
  sickness, 321
  teeth, 354
Lamb, William (1779–1848)
British statesman
  physician, 265
Lamont, Thomas W. (1870–1948)
American banker
  doctor, 112
Lamport, Felicia
  sinus, 324
Langer, Susan
  brain, 32
Lasker, Albert D. (1880–1952)
American advertising executive
  research, 306
Latham, Peter Mere
  chemistry, 41
  common sense, 50
  cure, 60
  diagnosis, 81
  disease, 92
  error, 134
  experience, 138
  experiment, 140
  facts, 143
  pain, 239
  physician, 265
  prescribe, 291
  remedy, 302
  symptom, 349
  theories, 357
  treatment, 365
Lederman, Leon (1922–)
US physicist
  prayer, 289
Legrain, G.
  heart, 165

Lerner, Max
  tranquilizers, 364
Lettsom, J.C. (1744–1815)
  physic, 248
Leviticus
  tooth, 353
Lewis, C.S. (1898–1963)
British scholar/writer
  death, 65
Lewis, Lucille
  nursing, 226
Lewis, Sinclair (1885–1951)
US novelist
  prayer, 289
Lippmann, Walter (1889–1974)
American newspaper commentator
  opinion, 236
Litin, Edward M.
  physician, 258
Livermore, Mary (1820–1905)
  physical organs, 251
Locke, John (1632–1704)
English political philosopher
  opinions, 236
Longfellow, Henry Wadsworth (1807–1882)
American poet
  cure, 60
  physician, 266
Lower, Lennie (1903–1947)
  dentist, 72
Luke
  physician, 267
Luttrell, Henry (1765–1851)
  sick, 317
Lyly, John (1554–1606)
English prose stylist
  medicine, 210

-M-

MacFadden, Bernarr (1868–1955)
  fever, 147
Mansfield, Katherine (1888–1923)
New Zealand short-story writer
  invalid, 183